Digital Sig
Instant Access

Digital Signal Processing: Instant Access

James D. Broesch
Dag Stranneby
William Walker

AMSTERDAM • BOSTON • HEIDELBERG • LONDON
NEW YORK • OXFORD • PARIS • SAN DIEGO
SAN FRANCISCO • SINGAPORE • SYDNEY • TOKYO
Newnes is an imprint of Elsevier

Newnes is an imprint of Elsevier
30 Corporate Drive, Suite 400, Burlington, MA 01803, USA
Linacre House, Jordan Hill, Oxford OX2 8DP, UK

 Recognizing the importance of preserving what has been written, Elsevier prints its
books on acid-free paper whenever possible.

Library of Congress Cataloguing-in-Publication Data
Application submitted.

British Library Cataloguing-in-Publication Data
A catalogue record for this book is available from the British Library.

ISBN: 978-0-7506-8976-2

For information on all Newnes publications
visit our Web site at: www.books.elsevier.com

Printed and bound by CPI Group (UK) Ltd, Croydon, CR0 4YY

Transferred to Digital Print 2011

Typeset by Charon Tec Ltd., A Macmillan Company (www.macmillansolutions.com)

Working together to grow
libraries in developing countries

www.elsevier.com | www.bookaid.org | www.sabre.org

ELSEVIER   BOOK AID
International   Sabre Foundation

Contents

Chapter 1

Why DSP?

In an Instant

- DSP Definitions
- The Need for DSP
- Learning Digital Signal Processing Techniques
- Instant Summary

DSP Definitions

The acronym *DSP* is used for two terms, *digital signal processing* and *digital signal processor*, both of which are covered in this book. *Digital signal processing* is performing signal processing using digital techniques with the aid of digital hardware and/or some kind of computing device. Signal processing can of course be analog as well, but, for a variety of reasons, we may prefer to handle the processing digitally. A digital computer or processor that is designed especially for signal processing applications is called a *digital signal processor*.

THE NEED FOR DSP

To understand the relative merits of analog and digital processing, it is convenient to compare the two techniques in a common application. Figure 1.1 shows two approaches to recording sounds such as music or speech. Figure 1.1a is the analog approach. It works like this:

- Sound waves impact the microphone, where they are converted to electrical impulses.
- These electrical signals are amplified, then converted to magnetic fields by the recording head.
- As the magnetic tape moves under the head, the intensity of the magnetic fields is stored on the tape.

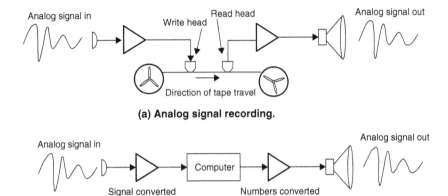

(a) Analog signal recording.

(b) Digital signal recording.

FIGURE 1.1   Analog and digital systems

The playback process is just the inverse of the recording process:

- As the magnetic tape moves under the playback head, the magnetic field on the tape is converted to an electrical signal.
- The signal is then amplified and sent to the speaker. The speaker converts the amplified signal back to sound waves.

The advantage of the analog process is twofold: first, it is conceptually quite simple. Second, by definition, an analog signal can take on virtually an infinite number of values within the signal's dynamic range. Unfortunately, this analog process is inherently unstable. The amplifiers are subject to gain variation over temperature, humidity, and time. The magnetic tape stretches and shrinks, thus distorting the recorded signal. The magnetic fields themselves will, over time, lose some of their strength. Variations in the speed of the motor driving the tape cause additional distortion. All of these factors combine to ensure that the output signal will be considerably lower in quality than the input signal Each time the signal is passed on to another analog process, these adverse effects are multiplied. It is rare for an analog system to be able to make more than two or three generations of copies.

Now let's look at the digital process as shown in Figure 1.1b:

- As in the analog case, the sound waves impact the microphone and are converted to electrical signals. These electrical signals are then amplified to a usable level.
- The electrical signals are measured or, in other words, they are converted to numbers.
- These numbers can now be stored or manipulated by a computer just as any other numbers are.

- To play back the signal, the numbers are simply converted back to electrical signals. As in the analog case, these signals are then used to drive a speaker.

There are two distinct disadvantages to the digital process: first, it is far more complicated than the analog process; second, computers can only handle numbers of finite resolution. Thus, the (potentially) "infinite resolution" of the analog signal is lost.

Insider Info

The first major contribution in the area of digital filter synthesis was made by Kaiser at Bell Laboratories. His work showed how to design useful filters using the bilinear transform. Further, in about 1965 the famous paper by Cooley and Turkey was published. In this paper, FFT (fast Fourier transform), an efficient and fast way of performing the DFT (discrete Fourier transform) was demonstrated.

Advantages of DSP

Obviously, there must be some compensating benefits of the digital process, and indeed there are. First, once converted to numbers, the signal is unconditionally stable. Using techniques such as error detection and correction, it is possible to store, transmit, and reproduce numbers with no corruption. The twentieth generation of recording is therefore just as accurate as the first generation.

Insider Info

The problems with analog signal reproduction have some interesting implications. Future generations will never really know what the Beatles sounded like, for example. The commercial analog technology of the 1960s was simply not able to accurately record and reproduce the signals. Several generations of analog signals were needed to reproduce the sound: First, a master tape would be recorded, and then mixed and edited; from this, a metal master record would be produced, from which would come a plastic impression. Each step of the process was a new generation of recording, and each generation acted on the signal like a filter, reducing the frequency content and skewing the phase. As with the paintings in the Sistine Chapel, the true colors and brilliance of the original art is lost to history. Things are different for today s musicians. A thousand years from now historians will be able to accurately play back the digitally mastered CDs of today. The discs themselves may well deteriorate, but before they do, the digital numbers on them can be copied with perfect accuracy. Signals stored digitally are really just large arrays of numbers. As such, they are immune to the physical limitations of analog signals.

There are other significant advantages to processing signals digitally. Geophysicists were one of the first groups to apply the techniques of signal processing. The seismic signals of interest to them are often of very low frequency, from 0.01 Hz to 10 Hz. It is difficult to build analog filters that work at these low frequencies. Component values must be so large that physically implementing the filter may well be impossible. Once the signals have been converted to digital numbers, however, it is a straightforward process to program a computer to perform the filtering.

Other advantages to digital signals abound. For example, DSP can allow large bandwidth signals to be sent over narrow bandwidth channels. A 20-kHz signal can be digitized and then sent over a 5-kHz channel. The signal may take four times as long to get through the narrower bandwidth channel, but when it comes out the other side it can be reconstructed to its full 20-kHz bandwidth.

In the same way, communications security can be greatly improved through DSP. Since the signal is sent as numbers, it can be easily encrypted. When received, the numbers are decrypted and then reproduced as the original signal. Modern "secure telephone" DSP systems allow this processing to be done with no detectable effect on the conversation.

Technology Trade-offs

DSP has several major advantages over analog signal processing techniques, including:

- Essentially perfect reproducibility
- Guaranteed accuracy (no individual tuning and pruning needed)
- Well-suited for volume production

LEARNING DIGITAL SIGNAL PROCESSING TECHNIQUES

The most important first step of studying any subject is to grasp the overall picture and to understand the basics before diving into the depths. With that in mind, the goal of this book is to provide a broad introduction and overview of DSP techniques and applications. The authors seek to bring an intuitive understanding of the concepts and systems involved in the field of DSP engineering.

Only a few years ago, DSP techniques were considered advanced and esoteric subjects, their use limited to research labs or advanced applications such as radar identification. Today, the technology has found its way into virtually every segment of electronics. Computer graphics, mobile entertainment and communication devices, and automobiles are just a few of the common examples.

The rapid acceptance and commercialization of this technology has presented the modern design engineer with a serious challenge: either gain a working knowledge of these techniques or risk obsolescence. Traditionally, engineers have had two options for acquiring new skills: go back to school, or turn to vendors' technical documentation. In the case of DSP, neither of these is a particularly good option.

Undergraduate programs—and even many graduate programs—devoted to DSP are really only thinly disguised courses in the mathematical discipline known as complex analysis. These programs do not aim to teach a working knowledge of DSP, but rather to prepare students for graduate research on DSP topics. Much of the information that is needed to comprehend the "whys and wherefores" of DSP are not covered.

Manufacturer documentation is often of little more use to the uninitiated. Application notes and design guides usually focus on particular features of the vendor's instruction set or architecture.

In this book, we hope to bridge the gap between the theory of DSP and the practical knowledge necessary to understand a working DSP system. The mathematics is not ignored; you will find many sophisticated mathematical relationships in thumbing through the pages of this book. What is left out, however, are the formal proofs, the esoteric discussions, and the tedious mathematical exercises. In their place are background discussions explaining how and why the math is important, examples to run on any general-purpose computer, and tips that can help you gain a comfortable understanding of the DSP processes.

INSTANT SUMMARY

- Digitally processing a signal allows us to do things with signals that would be difficult, or impossible, with analog approaches.
- With modern components and techniques, these advantages can often be realized economically and efficiently.

Chapter 2

The Analog-Digital Interface

In an Instant

- Definitions
- Sampling and Reconstruction
- Quantization
- Encoding and Modulation

- Number Representations
- Digital-to-Analog Conversion
- Analog-to-Digital Conversion
- Instant Summary

Definitions

In most systems, whether electronic, financial or social, the majority of problems arise in the interface between different subparts. This is also true for digital signal processing systems. Most signals in real life are continuous in amplitude and time—that is, *analog*—but our digital system is working with amplitude- and time-discrete signals, or so-called *digital* signals. So, the input signals entering our system need to be converted from analog to digital form before the actual signal processing can take place.

For the same reason, the output signals from our DSP device usually need to be reconverted back from digital to analog form, to be used in, for instance, hydraulic valves or loudspeakers or other analog actuators. These conversion processes between the analog and digital world also add some problems to our system. These matters will be addressed in this chapter, together with a brief presentation of some common techniques to perform the actual conversion processes.

First we will define some of the important terms encountered in this chapter. *Sampling* is the process of going from a continuous signal to a discrete signal. An *analog-to-digital converter* (ADC) is a device that converts an analog voltage into a digital number. There are a number of different types, but the most common ones used in DSP are the *successive approximation register* (SAR) and the *flash converter*. A *digital-to-analog converter* converts a digital number to an analog voltage. All of these terms will be further explained as we move through the material in this chapter.

SAMPLING AND RECONSTRUCTION

Recall that sampling is how we go from a continuous (analog) signal to a discrete (digital) signal. Sampling can be regarded as multiplying the time-continuous signal $g(t)$ with a train of unit pulses $p(t)$ (see Figure 2.1)

$$g^{\#}(t) = g(t)p(t) = \sum_{n=-\infty}^{+\infty} g(nT)\delta(t - nT) \qquad (2.1)$$

where $g^{\#}(t)$ is the sampled signal. Since the unit pulses are either one or zero, the multiplication can be regarded as a pure switching operation.

The time period T between the unit pulses in the pulse train is called the *sampling period*. In most cases, this period is constant, resulting in "equidistant sampling." In most systems today, it is common to use one or more constant sampling periods. The sampling period T is related to the *sampling rate* or *sampling frequency* f_s such that

$$f_s = \frac{\omega_s}{2\pi} = \frac{1}{T} \qquad (2.2)$$

Insider Info

The sampling period does not have to be constant. In some systems, many different sampling periods are used (called multirate sampling). *In other applications, the sampling period may be a stochastic variable, resulting in* random sampling, *which complicates the analysis considerably.*

The process of sampling implies reduction of knowledge. For the time-continuous signal, we know the value of the signal at every instant of time, but for the sampled version (the time-discrete signal) we only know the value at specific points in time. If we want to reconstruct the original time-continuous signal from the time-discrete sampled version, we have to make more or less qualified interpolations of the values in between the sampling points. If our interpolated values differ from the true signal, we have introduced distortion in our reconstructed signal.

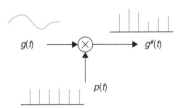

FIGURE 2.1   Sampling viewed as a multiplication process

If the sampling frequency is less than twice the maximum analog signal frequency, a phenomenon called *aliasing* will occur, which distorts the sampled signal. We will discuss aliasing in more detail in the next chapter.

Key Concept

In order to avoid aliasing distortion in the sampled signal, it is imperative that the bandwidth of the original time-continuous signal being sampled is smaller than half the sampling frequency (also called the Nyquist frequency).

To avoid aliasing distortion in practical cases, the sampling device is always preceded by some kind of low-pass filter (*antialiasing filter*) to reduce the bandwidth of the incoming signal. This signal is often quite complicated and may contain a large number of frequency components. Since it is impossible to build perfect filters, there is a risk of too-high-frequency components leaking into the sampler, causing aliasing distortion. We also have to be aware that high-frequency interference may somehow enter the signal path after the low-pass filter, and we may experience aliasing distortion even though the filter is adequate.

If the Nyquist criteria is met and hence no aliasing distortion is present, we can reconstruct the original bandwidth-limited, time-continuous signal $g(t)$ in an unambiguous way.

QUANTIZATION

The sampling process described in the previous section is the process of converting a continuous-time signal into a discrete-time signal, while *quantization* converts a signal continuous in amplitude into a signal discrete in amplitude.

Quantization can be thought of as classifying the level of the continuous-valued signal into certain bands. In most cases, these bands are equally spaced over a given range and undesired nonlinear band spacing may cause harmonic distortion.

Every band is assigned a code or numerical value. Once we have decided to which band the present signal level belongs, the corresponding code can be used to represent the signal level.

Most systems today use the binary code; i.e., the number of quantization intervals N are

$$N = 2^n \tag{2.3}$$

where n is the word length of the binary code. For example, with $n = 8$ bits we get a *resolution* of $N = 256$ bands, $n = 12$ yields $N = 4096$, and $n = 16$ gives $N = 65536$ bands. Obviously, the more bands we have—i.e., the longer the word length—the better resolution we obtain. This in turn renders a more accurate representation of the signal.

Insider Info

Another way of looking at resolution of a quantization process is to define the dynamic range as the ratio between the strongest and the weakest signal level that can be represented. The dynamic range is often expressed in decibels. Since every new bit of word length being added increases the number of bands by a factor of 2 the corresponding increase in dynamic range is 6 dB. Hence, an 8-bit system has a dynamic range of 48 dB, a 12-bit system has 72 dB, etc. (This of course only applies for linear band spacing.)

ENCODING AND MODULATION

Assuming we have converted our analog signals to numbers in the digital world, there are many ways to *encode* the digital information into the shape of electrical signals. This process is called *modulation*. The most common method is probably *pulse code modulation* (PCM). There are two common ways of transmitting PCM, and they are *parallel* and *serial* mode. In an example of the parallel case, the information is encoded as voltage levels on a number of wires, called a parallel *bus*. We are using binary signals, which means that only two voltage levels are used, +5 V corresponding to a binary "1" (or "true"), and 0 V meaning a binary "0" (or "false"). Hence, every wire carrying 0 or +5 V contributes a binary digit ("bit"). A parallel bus consisting of eight wires will therefore carry 8 bits, a byte consisting of bits D0, D1–D7 (Figure 2.2).

Technology Trade-offs

Parallel buses are able to transfer high information data rates, since an entire data word (a sampled value) is being transferred at a time. This transmission can take place between, for instance, an analog-to-digital converter (ADC) and a digital signal processor (DSP). One drawback with parallel buses is that they

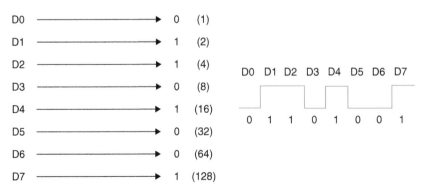

FIGURE 2.2   Example, a byte (96H) encoded (weights in parenthesis) using PCM in parallel mode (parallel bus, 8 bits, eight wires) and in serial mode as an 8-bit pulse train (over one wire)

require a number of wires, taking up more board space on a printed circuit board. Another problem is that we may experience skew problems, i.e. different time delays on different wires, meaning that all bits will not arrive at the same time in the receiver end of the bus, and data words will be messed up. Since this is especially true for long, high-speed parallel buses, this kind of bus is only suited for comparatively short transmission distances. Protecting long parallel buses from picking up wireless interference or radiating interference may also be a formidable problem. The alternative way of dealing with PCM signals is to use the *serial* transfer mode where the bits are transferred in sequence on a single wire (see Figure 2.2). Transmission times are longer, but only one wire is needed. Board space and skew problems will be eliminated and the interference problem can be easier to solve.

There are many possible modulation schemes, such as pulse amplitude modulation (PAM), pulse position modulation (PPM), pulse number modulation (PNM), pulse width modulation (PWM) and pulse density modulation (PDM). All these modulation types are used in serial transfer mode (see Figure 2.3).

- **Pulse amplitude modulation (PAM)** The actual amplitude of the pulse represents the number being transmitted. Hence, PAM is continuous in amplitude but discrete in time. The output of a sampling circuit with a zero-order hold (ZOH) is one example of a PAM signal.
- **Pulse position modulation (PPM)** A pulse of fixed width and amplitude is used to transmit the information. The actual number is represented by the position in time where the pulse appears in a given time slot.

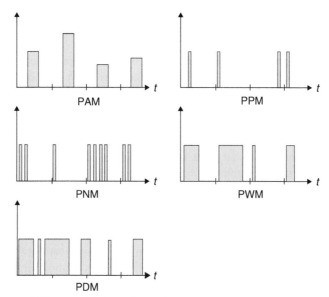

FIGURE 2.3   Different modulation schemes for serial mode data communication, PAM, PPM, PNM, PWM and PDM

- **Pulse number modulation (PNM)** Related to PPM in the sense that we are using pulses with fixed amplitude and width. In this modulation scheme, however, many pulses are transmitted in every time slot, and the number of pulses present in the slot represents the number being transmitted.
- **Pulse width modulation (PWM)** Quite common modulation scheme, especially in power control and power amplifier contexts. In this case, the width (duration) T_1 of a pulse in a given time slot T represents the number being transmitted.
- **Pulse density modulation (PDM)** May be viewed as a type of degenerated PWM, in the sense that not only the width of the pulses changes, but also the periodicity (frequency). The number being transmitted is represented by the density or "average" of the pulses.

Insider Info

Some signal converting and processing chips and subsystems may use different modulation methods to communicate. This may be due to standardization or due to the way the actual circuit works. One example is the so-called CODEC (coder–decoder). This is a chip used in telephone systems, containing both an analog-to-digital converter (ADC) and a digital-to-analog converter (DAC) and other necessary functions to implement a full two-way analog-digital interface for voice signals. Many such chips use a serial PCM interface. Switching devices and digital signal processors commonly have built-in interfaces to handle these types of signals.

NUMBER REPRESENTATION

When the analog signal is quantized, it is commonly represented by binary numbers in the following processing steps. There are many possible representations of quantized amplitude values. One way is to use *fixed-point* formats like *2's complement*, *offset binary* or *sign and magnitude*. Another way is to use some kind of *floating-point* format. The difference between the fixed-point formats can be seen in Table 2.1.

The most common fixed-point representation is 2's complement. In the digital signal processing community, we often interpret the numbers as *fractions* rather than integers. Other codes used are Gray code and binary-coded decimal (BCD).

There are a number of floating-point formats around. They all rely on the principle of representing a number in three parts: a sign bit, an exponent and a mantissa. One such common format is the Institute of Electrical and Electronics Engineers *(IEEE) Standard 754.1985* single precision 32-bit format, where the floating-point number is represented by one sign bit, an 8-bit exponent and a 23-bit mantissa. Using this method, numbers between $\pm 3.37 \cdot 10^{38}$ and $\pm 8.4 \cdot 10^{-37}$ can be represented using only 32 bits.

TABLE 2.1 Some fixed-point binary number formats

Integer	2's complement	Offset binary	Sign and magnitude
7	0111	1111	0111
6	0110	1110	0110
5	0101	1101	0101
4	0100	1100	0100
3	0011	1011	0011
2	0010	1010	0010
1	0001	1001	0001
0	0000	1000	0000
−1	1111	0111	1000
−2	1110	0110	1001
−3	1101	0101	1010
−4	1100	0100	1011
−5	1011	0011	1100
−6	1010	0010	1101
−7	1001	0001	1110
−8	1000	0000	1111

Technology Trade-offs

Note that the use of floating-point representation expands the dynamic range but at the expense of the resolution and system complexity. For instance, a 32-bit fixed-point system may have better resolution than a 32-bit floating-point system, since in the floating-point case, the resolution is determined by the word length of the mantissa being only 23 bits. Another problem with floating-point systems is the *signal-to-noise ratio* (SNR). Since the size of the quantization steps will change as the exponent changes, so will the quantization noise. Hence, there will be discontinuous changes in SNR at specific signal levels. In an audio system, audible distortion may result from the modulation and quantization noise created by barely audible low-frequency signals causing numerous exponent switches.

From the above, we realize that fixed-point (linear) systems yield *uniform* quantization of the signal, while floating-point systems, due to the range changing, provide a *nonuniform* quantization.

Technology Trade-offs

Nonuniform quantization is often used in systems where a compromise between word length, dynamic range and distortion at low signal levels has to be found. By using larger quantization steps for larger signal levels and smaller steps for weak signals, a good dynamic range can be obtained without causing serious distortion at low signal levels or requiring unreasonable word lengths (number of quantization steps).

DIGITAL-TO-ANALOG CONVERSION

The task of the digital-to-analog converter (DAC) is to convert a numerical, commonly binary digital value into an analog output signal. The DAC is subject to many requirements, such as offset, gain, linearity, monotonicity and settling time. Several of the most important of these requirements are defined here:

Offset is the analog output when the digital input calls for a zero output. This should of course ideally be zero. The offset error affects all output signals with the same additive amount and in most cases it can be sufficiently compensated for by external circuits or by *trimming* the DAC.

Gain or scale factor is the slope of the transfer curve from digital numbers to analog levels. Hence, the gain error is the error in the slope of the transfer curve. This error affects all output signals by the same percentage amount, and can normally be (almost) eliminated by trimming the DAC or by means of external circuitry.

Linearity can be subdivided into integral linearity (relative accuracy) and differential linearity. *Integral linearity* error is the deviation of the transfer curve from a straight line (the output of a perfect DAC). This error is not possible to adjust or compensate for easily. *Differential linearity* measures the difference between any two adjacent output levels minus the step size for one LSB. If the output level for one step differs from the previous step by exactly the value corresponding to one least significant bit (LSB) of the digital value, the differential nonlinearity is zero. Differential linearity errors cannot be eliminated easily.

Monotonicity implies that the analog output must increase as the digital input increases, and decrease as the input decreases for all values over the specified signal range. Non-monotonicity is a result of excess differential non-linearity (≥ 1 LSB). Monotonicity is essential in many control applications to maintain precision and to avoid instabilities in feedback loops.

Absolute accuracy error is the difference between the measured analog output from a DAC compared to the expected output for a given digital input. The absolute accuracy is the compound effect of the offset error, gain error and linearity errors described above.

Settling time of a DAC is the time required for the output to approach a final value within the limits of an allowed error band for a step change in the digital input. Measuring the settling time may be difficult in practice, since some DACs produce glitches when switching from one level to another.

DAC settling time is a parameter of importance mainly in high sampling rate applications.

> **Alert!**
>
> **One important thing to remember is that these parameters may be affected by supply voltage and temperature. In DAC data sheets, the parameters are only specified for certain temperatures and supply voltages, such as normal room temperature +25°C and nominal supply voltage. Considerable deviations from the specified figures may occur in a practical system.**

Types of DACs

Several types of DACs are used in DSP systems, as described in the following sections:

Multiplying DACs

This is the most common form of DAC. The output is the product of an input current or reference voltage and an input digital code. The digital information is assumed to be in PCM parallel format. There are also DACs with a built-in shift register circuit, converting serial PCM to parallel. Hence, there are multiplying DACs for both parallel and serial transfer mode PCM available. Multiplying DACs have the advantage of being fast.

In Figure 2.4 a generic current source multiplying DAC is shown. The bits in the input digital code are used to turn on a selection of current sources, which are then summed to obtain the output current. The output current can easily be converted into an output voltage using an operational amplifier. There are many design techniques used to build this type of DAC, including R-2R ladders and charge redistribution techniques.

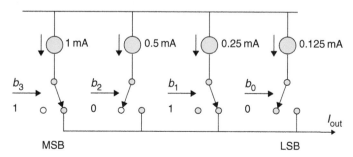

FIGURE 2.4 A generic multiplying DAC using current sources controlled by the bits in the digital input code

Integrating DACs

This class of DACs is also called *counting* DACs. These DACs are often slow compared to the multiplying converters. On the other hand, they may offer high resolution using quite simple circuit elements.

The basic building blocks are: an "analog accumulator" usually called an integrator, a voltage (or current) reference source, an analog selector, a digital counter and a digital comparator. Figure 2.5 shows an example of an integrating DAC. The incoming N-bits PCM data (parallel transfer mode) is fed to one input of the digital comparator. The other input of the comparator is connected to the binary counter having N bits, counting pulses from a clock oscillator running at frequency f_c

$$f_c \geq 2^N f_s \qquad (2.4)$$

where f_s is the sampling frequency of the system and N is the word length.

The integrator simply averages the PWM signal presented to the input, thus producing the output voltage U_{out}. The precision of this DAC depends on the stability of the reference voltages, the performance of the integrator and the timing precision of the digital parts, including the analog selector.

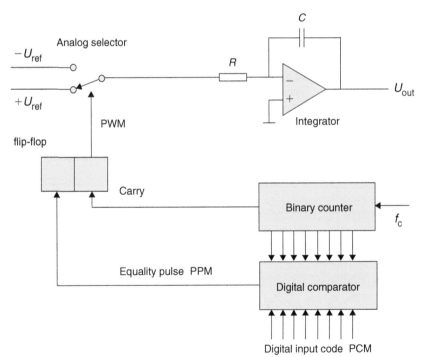

FIGURE 2.5   An example integrating (counting) DAC. (In a real world implementation, additional control and synchronization circuits are needed.)

Bitstream DACs

This type of DAC relies on the *oversampling* principle—that is, using a considerably higher sampling rate than that required by the Nyquist criteria. Using this method, sampling rate can be traded for accuracy of the analog hardware and the requirements of the analog reconstruction filter on the output can be relaxed. Oversampling reduces the problem of accurate N-bit data conversion to a rapid succession of, for instance, 1-bit D/A conversions. Since the latter operation involves only an analog switch and a reference voltage source, it can be performed with high accuracy and linearity.

The concept of oversampling is to increase a fairly low sampling frequency to a higher one by a factor called the oversampling ratio (OSR). Increasing the sampling rate implies that more samples are needed than are available in the original data stream. Hence, "new" sample points in between the original ones have to be created. This is done by means of an *interpolator*, also called an *oversampling filter*. The simplest form of interpolator creates new samples by making a linear interpolation between two "real" samples. In many systems, more elaborate interpolation functions are often used, implemented as a cascade of digital filters. As an example, an oversampling filter in a CD player may have 16-bit input samples at 44.1 kHz sampling frequency and an output of 28-bit samples at 176.4 kHz, i.e., an OSR of 4.

The interpolator is followed by the *truncator* or *M*-bit *quantizer*. The task of the truncator is to reduce the number of N bits in the incoming data stream to M bits in the outgoing data stream $(N > M)$.

Sample-and-Hold and Reconstruction Filters

The output from a DAC can be regarded as a PAM representation of the digital signal at the sampling rate. An ideal sample represents the value of the corresponding analog signal in a single point in time. Hence, in an *ideal* case, the output of a DAC is a train of impulses, each having an infinitesimal width, thus eliminating any *aperture error*. The aperture error is caused by the fact that a sample in a practical case does occupy a certain interval of time. The narrower the pulse width of the sample, the less the error. Of course, ideal DACs cannot be built in practice.

Technology Trade-offs

Another problem with real world DACs is that during the transition from one sample value to another, glitches, ringing and other types of interference may occur. To counteract this, a *sample-and-hold* device (S&H or S/H) is used. The most common type is the *zero-order hold* (ZOH). This device keeps the output constant until the DAC has settled on the next sample value. Hence, the output of the S/H is a staircase waveform approximation of the sampled analog signal. In many cases, the S/H is built into the DAC itself.

In many cases an analog *reconstruction filter* or *smoothing filter* (or anti-image filter) is needed in the signal path after the S/H (see Technology Trade-offs) to achieve a good enough reconstruction of the analog signal. Since the filter must be implemented using analog components, it tends to be bulky and expensive and it is preferably kept simple, of the order of 3 or lower. A good way of relaxing the requirements of the filter is to use oversampling as described above. There are also additional requirements on the reconstruction filter depending on the application. In a high-quality audio system, there may be requirements regarding linear-phase shift and transient response, while in a feedback control system time delay parameters may be crucial.

ANALOG-TO-DIGITAL CONVERSION

The task of the analog-to-digital converter (ADC) is the inverse of the digital-to-analog converter: to convert an analog input signal into a numerical digital value. The specifications for an ADC are similar to those for a DAC: offset, gain, linearity, missing codes, conversion time and so on, as explained below.

Offset error is the difference between the analog input level which causes a first bit transition to occur and the level corresponding to 1/2 LSB. This should of course ideally be zero, i.e., the first bit transition should take place at a level representing exactly 1/2 LSB. The offset error affects all output codes with the same additive amount and can in most cases be sufficiently compensated for by adding an analog DC level to the input signal and/or by adding a fixed constant to the digital output.

Gain or scale factor is the slope of the transfer curve from analog levels to digital numbers. Hence, the gain error is the error in the slope of the transfer curve. It affects all output codes by the same percentage amount, and can normally be counteracted by amplification or attenuation of the analog input signal. Compensation can also be done by multiplying the digital number with a fixed gain calibration constant.

Linearity can be subdivided into integral linearity (relative accuracy) and differential linearity. *Integral linearity* error is the deviation of code mid-points of the transfer curve from a straight line. This error is not possible to adjust or compensate for easily. *Differential linearity* measures the difference between input levels corresponding to any two adjacent digital codes. If the input level for one step differs from the previous step by exactly the value corresponding to one least significant bit (LSB), the differential nonlinearity is zero. Differential linearity errors cannot be eliminated easily.

Monotonicity implies that increasing the analog input level never results in a decrease of the digital output code. Nonmonotonicity may cause stability problems in feedback controls systems.

Missing codes in an ADC means that some digital codes can never be generated. It indicates that differential nonlinearity is larger than 1 LSB. The problem of missing codes is generally caused by a nonmonotonic behavior of the internal DAC.

Absolute accuracy error is the difference between the actual analog input to an ADC compared to the expected input level for a given digital output. The absolute accuracy is the compound effect of the offset error, gain error and linearity errors described above.

Conversion time of an ADC is the time required by the ADC to perform a complete conversion process. The conversion is commonly started by a "strobe" or synchronization signal, controlling the sampling rate.

> **Alert!**
>
> As with DACs, it is important to remember that the parameters above may be affected by supply voltage and temperature. Data sheets only specify the parameters for certain temperatures and supply voltages. Significant deviations from the specified figures may therefore occur in a practical system.

Types of ADCs

As with DACs, there are several different types of ADCs used in digital signal processing.

Flash ADCs

Flash type (or parallel) ADCs are the fastest due to their short conversion time and can therefore be used for high sampling rates. Hundreds of megahertz is common today. On the other hand, these converters are quite complex, they have limited word length and hence resolution (10 bits or less), they are quite expensive and often suffer from considerable power dissipation.

The block diagram of a simple 2-bit flash ADC is shown in Figure 2.6. The analog input is passed to a number of analog level comparators in parallel (i.e., a bank of fast operational amplifiers with high gain and low offset).

If the analog input level U_{in} on the positive input of a comparator is greater than the level of the negative input, the output will be a digital "one". Otherwise, the comparator outputs a digital "zero". Now, a reference voltage U_{ref} is fed to the voltage divider chain, thus obtaining a number of reference levels

$$U_k = \frac{k}{2^N} U_{ref} \tag{2.5}$$

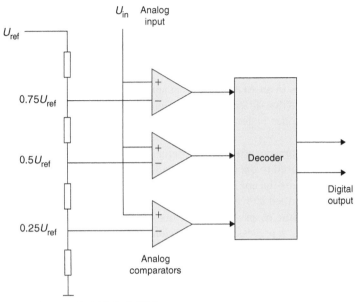

FIGURE 2.6   An example 2-bit flash ADC

where k is the quantization threshold number and N is the word length of the ADC. The analog input voltage will hence be compared to all possible quantization levels at the same time, rendering a "thermometer" output of digital ones and zeros from the comparators. These ones and zeros are then used by a digital decoder circuit to generate digital parallel PCM data on the output of the ADC.

As pointed out above, this type of ADC is fast, but is difficult to build for large word lengths. The resistors in the voltage divider chain have to be manufactured with high precision and the number of comparators and the complexity of the decoder circuit grows fast as the number of bits is increased.

SUCCESSIVE APPROXIMATION ADCS

These ADCs, also called successive approximation register (SAR) converters, are the most common ones today. They are quite fast, but not as fast as flash converters. On the other hand, they are easy to build and inexpensive, even for larger word lengths.

The main parts of the ADC are: an analog comparator, a digital register, a DAC and some digital control logic (see Figure 2.7). Using the analog comparator, the unknown input voltage U_{in} is compared to a voltage U_{DAC} created by a DAC, being a part of the ADC. If the input voltage is greater than the voltage coming from the DAC, the output of the comparator is a logic "one", otherwise a logic "zero". The DAC is fed an input digital code from the register, which is in turn controlled by the control logic. Now, the principle of successive approximation works as follows.

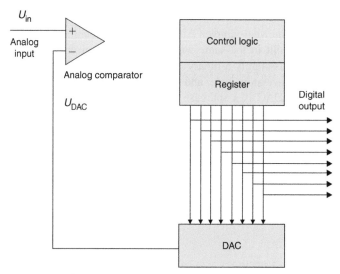

FIGURE 2.7   An example SAR ADC (simplified block diagram)

Assume that the register contains all zeros to start with, hence, the output of the DAC is $U_{DAC} = 0$. Now, the control logic will start to toggle the MSB to a one, and the analog voltage coming from the DAC will be half of the maximum possible output voltage. The control logic circuitry samples the signal coming from the comparator. If this is a one, the control logic knows that the input voltage is still larger than the voltage coming from the DAC and the "one" in the MSB will be left as is. If, on the other hand, the output of the comparator has turned zero, the output from the DAC is larger than the input voltage. Obviously, toggling the MSB to a one was just too much, and the bit is toggled back to zero. Now, the process is repeated for the second most significant bit and so on until all bits in the register have been toggled and set to a one or zero.

Hence, the SAR ADC always has a constant conversion time. It requires n approximation cycles, where n is the word length, i.e., the number of bits in the digital code. SAR-type converters of today may be used for sampling rates up to some megahertz.

Insider Info

An alternative way of looking at the SAR converter is to see it as a DAC + register put in a control feedback loop. We try to "tune" the register to match the analog input signal by observing the error signal from the comparator. Note that the DAC can of course be built in a variety of ways (see previous sections). Today, charge redistribution-based devices are quite common, since they are straightforward to implement using CMOS technology.

Counting ADCs

An alternative, somewhat simpler ADC type is the *counting* ADC. The converter is mainly built in the same way as the SAR converter but the control
logic and register is simply a binary counter. The counter is reset to all zeros
at the start of a conversion cycle, and is then incremented step by step. Hence,
the output of the DAC is a staircase ramp function. The counting maintains
until the comparator output switches to a zero and the counting is stopped.

The conversion time of this type of ADC depends on the input voltage; the
higher the level, the longer the conversion time (counting is assumed to take
place at a constant rate). The interesting thing about this converter is that the
output signal of the comparator is a PWM representation of the analog input
signal. Further, by connecting an edge-triggered, monostable flip-flop to the
comparator output, a PPM representation can also be obtained.

Integrating ADCs

Integrating ADCs (sometimes also called counting converters) are often quite
slow, but inexpensive and accurate. A common application is digital multime-
ters and similar equipment, in which precision and cost are more important
than speed.

There are many different variations of the integrating ADC, but the main
idea is that the unknown analog voltage (or current) is fed to the input of an
analog integrator with a well-known integration time constant $\tau = RC$. The
slope of the ramp on the output of the integrator is measured by taking the
time between the output level passing two or more fixed reference threshold
levels. The time needed for the ramp to go from one threshold to the other
is measured by starting and stopping a binary counter running at a constant
speed. The output of the counter is hence a measure of the slope of the integrator output, which in turn is proportional to the analog input signal level. Since
this type of ADC commonly has a quite long conversion time, i.e. integration
time, the input signal is required to be stable or only slowly varying. On the
other hand, the integration process will act as a low-pass filter, averaging the
input signal and hence suppressing interference superimposed on the analog
input signal to a certain extent.

Figure 2.8 shows a diagram of a simplified integrating ADC.

Sigma–delta ADCs

The *sigma–delta* ADC, sometimes also called a *bitstream* ADC, utilizes the
technique of oversampling, discussed earlier. One of the major advantages of
the sigma–delta ADC using oversampling is that it is able to use digital filtering and relaxes the demands on the analog anti-aliasing filter. This also implies
that about 90% of the die area is purely digital, cutting production costs.
Another advantage of using oversampling is that the quantization noise power
is spread evenly over a larger frequency spectrum than the frequency band of

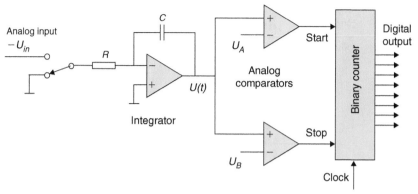

FIGURE 2.8   A simplified integrating ADC

interest. Hence, the quantization noise power in the signal band is lower than in the case of traditional sampling based on the Nyquist criteria.

Now, let us take a look at a simple 1-bit sigma–delta ADC. The converter uses a method that was derived from the delta modulation technique. This is based on quantizing the *difference* between successive samples, rather the quantizing the absolute value of the samples themselves. Figure 2.9 shows a *delta modulator* and demodulator with the modulator working as follows. From the analog input signal $x(t)$ a locally generated estimate $\tilde{x}(t)$ is subtracted. The difference $\varepsilon(t)$ between the two is fed to a 1-bit quantizer. In this simplified case, the quantizer may simply be the sign function, i.e. when $\varepsilon(t) \geq 0$ $y(n) = 1$, else $y(n) = 0$. The quantizer is working at the oversampling frequency, i.e. considerably faster than required by the signal bandwidth. Hence, the 1-bit digital output $y(n)$ can be interpreted as a kind of digital error signal:

$y(n) = 1$: estimated input signal level too small, increase level

$y(n) = 0$: estimated input signal level too large, decrease level

Now, the analog integrator situated in the feedback loop of the delta modulator (DM) is designed to function in exactly this way. Hence, if the analog input signal $x(t)$ is held at a constant level, the digital output $y(n)$ will (after

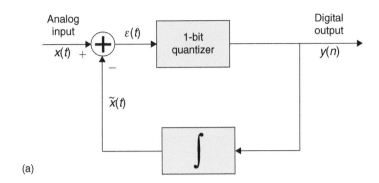

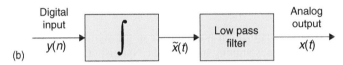

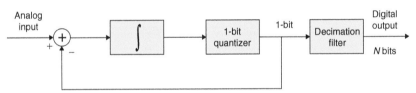

FIGURE 2.9   A simplified (a) delta modulator and (b) demodulator

FIGURE 2.10   A simplified, oversampled bitstream sigma–delta ADC

convergence) be a symmetrical square wave (0 1 0 1...), i.e. decrease, increase, decrease, increase... a kind of stable limit oscillation.

The delta demodulator is shown in the lower portion of Figure 2.9. The function is straightforward. Using the digital 1-bit "increase/decrease" signal, the estimated input level $\tilde{x}(t)$ can be created using an analog integrator of the same type as in the modulator. The output low-pass filter will suppress the ripple caused by the increase/decrease process.

Since integration is a linear process, the integrator in the demodulator can be moved to the input of the modulator. Hence, the demodulator will now only consist of the low-pass filter. We now have similar integrators on both inputs of the summation point in the modulator. For linearity reasons, these two integrators can be replaced by one integrator having $\varepsilon(t)$ connected to its input, and the output connected to the input of the 1-bit quantizer. The delta modulator has now become a sigma–delta modulator. The name is derived from the summation point (sigma) followed by the delta modulator.

If we now combine the oversampled 1-bit sigma–delta modulator with a digital *decimation filter* (rate reduction filter) we obtain a basic sigma–delta ADC (see Figure 2.10). The task of the decimation filter is threefold: to reduce

TABLE 2.2 Decimation filter example

Input	0 0 1 1 0	0 1 1 1 0	1 1 1 0 1	1 0 1 0 1	0 0 0 1 1
Averaging process	3×0; 2×1	2×0; 3×1	1×0; 4×1	2×0; 3×1	3×0; 2×1
Output	0	1	1	1	0

the sampling frequency, to increase the word length from 1 bit to N bits and to reduce any noise pushed back into the frequency range of interest by the crude 1-bit modulator. A simple illustration of a decimation filter, decimating by a factor 5, would be an averaging process as shown in Table 2.2.

INSTANT SUMMARY

In this chapter the following topics have been treated:

- Sampling and reconstruction
- Quantization
- Modulation: PCM, PAM, PPM, PNM, PWM and PDM
- Fixed-point 2's complement, offset binary, sign and magnitude and floating-point
- Multiplying, integrating and bitstream D/A converters
- Oversampling and interpolators
- Flash, successive approximation, counting and integrating A/D converters
- Sigma–delta and bitstream A/D converters

DSP System General Model

In an Instant

- Definitions
- The Big Picture
- Signal Acquisition
- More on Sampling Theory
- Sampling Resolution
- Instant Summary

Definitions

We have covered some details of analog and digital interfacing in the last chapter. Now let's now put together an entire DSP system model. First, we will define some terms you will encounter in this chapter. An *anti-aliasing filter* limits the bandwidth of any incoming signal to the system. A *smoothing filter* is used on the output of the DAC in a DSP system, to smooth out the stairstep pattern of the DAC's output. *Harvard architecture* is a common architecture for DSP processors; it splits the data path and the instruction path into two separate streams.

THE BIG PICTURE

The general model for a DSP system is shown in Figure 3.1. From a high-level point of view, a DSP system performs the following operations:

- Accepts an analog signal as an input.
- Converts this analog signal to numbers.
- Performs computations using the numbers.
- Converts the results of the computations back into an analog signal.

Optionally, different types of information can be derived from the numbers used in this process. This information may be analyzed, stored, displayed, transmitted, or otherwise manipulated.

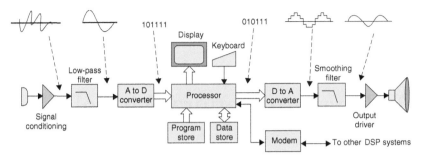

FIGURE 3.1   The general model for a DSP system

> **Key Concept**
>
> This model can be rearranged in several ways. For example, a CD player will not have the analog input section. A laboratory instrument may not have the analog output. The truly amazing thing about DSP systems, however, is that the model will fit *any* DSP application. The system could be a sonar or radar system, voice-mail system, video camera, or a host of other applications. The specifications of the individual key elements may change, but their function will remain the same.

In order to understand the overall DSP system, let's begin with a qualitative discussion of the key elements.

Input

All signal processing begins with an *input transducer*. The input transducer takes the input signal and converts it to an electrical signal. In signal-processing applications, the transducer can take many forms. A common example of an input transducer is a microphone. Other examples are geophones for seismic work, radar antennas, and infrared sensors. Generally, the output of the transducer is quite small: a few microvolts to several millivolts.

Signal-conditioning Circuit

The purpose of the signal-conditioning circuit is to take the few millivolts of output from the input transducer and convert it to levels usable by the following stages. Generally, this means amplifying the signal to somewhere between 3 and 12 V. The signal-conditioning circuit also limits the input signal to prevent damage to following stages. In some circuits, the conditioning circuit provides isolation between the transducer and the rest of the system circuitry.

Typically, signal-conditioning circuits are based on operational amplifiers or instrumentation amplifiers.

Anti-aliasing Filter

As mentioned in Chapter 2, the anti-aliasing filter is a low-pass filter, ideally having a flat passband and extremely sharp cutoff at the Nyquist frequency. Of course, building such a filter in practice is difficult and compromises have to be made. From a conceptual point of view, the anti-aliasing filter can be thought of as a mechanism to limit how fast the input signal can change. This is a critical function; the anti-aliasing filter ensures that the rest of the system will be able to track the signal. If the signal changes too rapidly, the rest of the system could miss critical parts of the signal.

Technology Trade-offs

Depending on the application, different requirements on the filter may be stated. In audio systems, linear phase response may, for example, be an important parameter, while in a DC-voltmeter instrument, a low offset voltage may be imperative. Designing proper anti-aliasing filters is typically not a trivial task, particularly if practical limitations such as board space and cost also have to be taken into account.

Analog-to-Digital Converter

As the name implies, the purpose of the analog-to-digital converter (ADC) is to convert the signal from its analog form to a digital data representation. Chapter 2 discusses this device in more detail. Due to the physics of converter circuitry, most ADCs require inputs of at least several volts for their full range input. Two of the most important characteristics of an ADC are the *conversion rate* and the *resolution*. The conversion rate defines how fast the ADC can convert an analog value to a digital value. The resolution defines how close the digital number is to the actual analog value.

The output of the ADC is a binary number that can be manipulated mathematically.

Processor

Theoretically, there is nothing special about the processor. It simply performs the calculations required for processing the signal. For example, if our DSP system is a simple amplifier, then the input value is literally multiplied by the gain (amplification) constant.

In the early days of signal processing, the processor was often a general-purpose mainframe computer. As the field of DSP progressed, special high-speed processors were designed to handle the "number crunching."

Today, a wide variety of specialized processors are dedicated to DSP These processors are designed to achieve very high data throughputs, using a combination of high-speed hardware, specialized architectures, and dedicated instruction sets. All of these functions are designed to efficiently implement DSP algorithms.

Technology Trade-offs

There are four different ways of implementing the required processor hardware:

- Conventional microprocessor
- DSP chip
- Bitslice or wordslice approach
- Dedicated hardware, FPGA (field programmable gate array), ASIC (application specific integrated circuit)

(These will be discussed in more detail in Chapter 8.)

Program Store, Data Store

The *program store* stores the instructions used in implementing the required DSP algorithms. In a general-purpose computer (von Neumann architecture), data and instructions are stored together. In most DSP systems, the program is stored separately from the data, since this allows faster execution of the instructions. Data can be moved on its own bus at the same time that instructions are being fetched. This architecture was developed from basic research performed at Harvard University, and therefore is generally called a *Harvard architecture*. Often the data bus and the instruction bus have different widths.

> **Insider Info**
>
> *Quite often, three system buses can be found on DSP systems, one for instructions, one for data (including I/O) and one for transferring coefficients from a separate memory area or chip.*

Data Transmission

DSP data is commonly transmitted to other DSP systems. Sometimes the data is stored in bulk form on magnetic tape, CDs, or other media. This ability to store and transmit the data in digital form is one of the key benefits of DSP operations. An analog signal, no matter how it is stored, will immediately begin to degrade. A digital signal, however, is much more robust since it is composed of ones and zeroes. Furthermore, the digital signal can be protected with error detection and correction codes.

Display and User Input

Not all DSP systems have displays or user input. However, it is often handy to have some visual representation of the signal. If the purpose of the system is to manipulate the signal, then obviously the user needs a way to input commands to the system. This can be accomplished with a specialized keypad, a few discrete switches, or a full keyboard.

Digital-to-Analog Converter

In many DSP systems, the signal must be converted back to analog form after it has been processed. This is the function of the digital-to-analog converter (DAC), as discussed in more detail in Chapter 2. Conceptually, DACs are quite straightforward: a binary number put on the input causes a corresponding voltage on the output. One of the key specifications of the DAC is how fast the output voltage settles to the commanded value. The slew rate of the DAC should be matched to the acquisition rate of the ADC.

Output Smoothing Filter

As the name implies, the purpose of the smoothing filter is to take the edges off the waveform coming from the DAC. This device was also discussed briefly in Chapter 2. This filter is necessary since the waveform will have a "stair-step" shape, resulting from the sequence of discrete inputs applied to the DAC. Generally, the smoothing filter is a simple low-pass system. Often, a basic RC circuit does the job.

Output Amplifier

The output amplifier is generally a straightforward amplifier with two main purposes. First, it matches the high impedance of the DAC to the low imped-ance of the transducer. Second, it boosts the power to the level required.

Output Transducer

Like the input transducer, the output transducer can assume a variety of forms. Common examples are speakers, antennas, and motors.

SIGNAL ACQUISITION

In most practical DSP applications, we will be acquiring a signal and then doing some manipulation on this signal. This work is often called *digital signal analysis.*

One of the first things we must do when we are designing a system to han-dle a signal is to determine what performance is required. In other words, *how do we know that our system can handle the signal?* The answer to this ques-tion, naturally, involves a number of issues. Some of the issues are the same ones that we would deal with when designing any system:

- Are the voltages coming into our system within safe ranges?
- Will our design provide adequate bandwidth to handle the signal?
- Is there enough power to run the equipment?
- Is there enough room for the hardware?

We must also consider some additional requirements that are specific to DSP systems or are strongly influenced by the fact that the signals will be handled digitally. These include:

- How many samples per second will be required to handle the signal?
- How much resolution is required to process the signal accurately?
- How much of the signal will need to be kept in memory?
- How many operations must we do on each sample of the signal?

Stating the requirements in general terms is straightforward. We must ensure that the incoming analog signal is sufficiently bandwidth-limited for our system to handle it; the number of samples per second must be sufficient to accurately represent the analog signal in digital form; the resolution must be sufficient to ensure that the signal is not distorted beyond acceptable limits; and our system must be fast enough to do all required calculations.

Obviously, however, these are qualitative requirements. To determine these requirements explicitly requires both theoretical understanding and practical knowledge of how a DSP system works. In the next section we will look at one of the major design requirements: the number of samples per second.

MORE ON SAMPLING THEORY

As we learned earlier, one important question to ask is: *what is the maximum frequency we can handle for a given number of samples per second?* We can get a good feeling for the answer by looking at an example. The graph shown by the dashed line in Figure 3.2 is an 8-Hz analog signal, but if we sampled this signal at a sampling frequency of 16 samples/sec. then we would get a DC value of zero. What went wrong?

In Chapter 2, we learned the importance of the Nyquist frequency, and we will restate it here.

Key Concept

The frequency of our sampled signal must be less than half the number of samples per second.

This is a key building block in what is known as the *Nyquist theorem.* We do not yet have all of the pieces to present a discussion of the Nyquist theorem, but we will shortly.

In the meantime, let's explore the significance of our discovery a little further. Clearly, this is a manifestation of the difference between the analog frequency and the digital frequency. Intuitively, we can think of it as follows: To represent one cycle of a sine wave, what are the minimum number of points needed? For most cases, any two points are adequate. If we know that any two separate points are points on *one* cycle of a sine wave, we can fit a curve to the sine wave. There is one important exception to this, however: when the two points have a value of zero. We need *more than* two points per cycle to ensure that we can accurately produce the desired waveform.

From Figure 3.2, we can see that we get the same output for a frequency *f* of either 0 or 8 when we are using 16 samples/second. For this reason, these frequencies are said to be *aliases* of one another.

We just "proved," in a nonrigorous way, that our maximum digital frequency is N/2. But what happens if we were to put in values for *f* greater than N/2? For example, what if we put in a value of, say, 10 for *f* when N = 16? The answer is that it will alias to a value of 2, just as a value of 8 aliased to a value of 0. If we keep playing at this, we soon see that we can only generate output frequencies for a range of 0 to N/2.

We define digital frequency as $\lambda = \omega T$. If we substitute N/2 for *f* and expand this we get:

$$\begin{aligned}
\lambda &= \omega T \\
&= 2\pi f T \\
&= 2\pi \left(\frac{N}{2} \right) \frac{1}{N} \\
&= \pi
\end{aligned}$$

(3.1)

It would therefore appear that our digital frequency must be between 0 and π. We can use any other value we want, but if it is outside this range, it will map to a frequency that is within the range of 0 to π. However, note that we said it would "*appear* that our digital frequency must be between 0 and π." This is because we haven't quite covered all of the bases.

Normally, in electronics we don't think of frequency as having a sign. However, negative frequencies are possible in the real world and are no great mystery. It simply means that the phase between the real and imaginary components are opposite what they would be for a positive frequency. In the case of a point on the unit circle, a negative frequency means that the point is rotating clockwise rather than counterclockwise. The sign of the frequency for a purely real or a purely imaginary signal is meaningful only if there is some way to reference the phase. (We will discuss real and imaginary signals in Chapter 4, on the mathematics of DSP.)

The signals generated so far have been real, but there is no reason not to plug in a negative value of *f*. Since $\sin(-\omega) = -\sin(\omega)$, we would get the same frequency out, but it would be 180° out of phase. Still, this phase difference does make the signal unique; thus, the actual unique range of a digital frequency is $-\pi$ to π.

This discussion may seem a bit esoteric, but it definitely has practical significance. A common practice is to specify the performance of a DSP *algorithm* over the range of $-\pi$ to π. The DSP *system* will map this range to analog frequencies by selection of the number of samples per second.

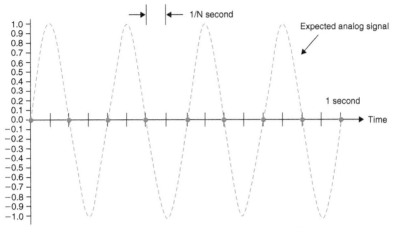

8-Hz signal generated with 16 samples/second. Actual digital signal is DC at 0 V.

FIGURE 3.2   Aliasing

The second part of demonstrating the Nyquist theorem lies in showing that what is true for sine waves will, if we are careful, apply to any waveform. We will do this in the section covering the Fourier series.

SAMPLING RESOLUTION

In order to generate, capture, or reproduce a real-world analog signal, we must ensure that we represent the signal with sufficient resolution. Generally, resolution will have two characteristics:

- The number of samples per second.
- The resolution of the amplitude of each sample.

The resolution of the amplitude of each sample is a system parameter. In other words, it will depend upon the input circuitry, how the system is used, and so forth. However, the theoretical limit for the amplitude resolution is defined by the number of bits resolved in the ADC or converted by the DAC.

The formula for determining the resolution of a system is:

$$r_{min} = \frac{1}{2^n - 1} \tag{3.2}$$

where n is the number of bits. For example, if we have a 2-bit system, then the maximum resolution will be:

$$r_{min} = \frac{1}{3} 1$$

Looking at this in table form (Table 3.1) shows the mapping for each of the possible binary values.

TABLE 3.1 Binary mapping

Binary value	Weight
00	0
01	1/3
10	2/3
11	1

Notice that we have expressed the *weight* for each possible binary value. As with the case of digital versus analog frequency, we can only express the digital value as a dimensionless number. The actual amplitude depends on the scaling performed by the DAC or the ADC. Notice that in this example we are dealing with only positive values. In practice there are a number of different schemes for setting weights. Twos complement and offset binary are two of the most common schemes used in signal processing.

Let's look at a typical example. Assume that we are designing a system to monitor an important parameter in a control system. The signal has a possible range of -5 volts to $+5$ volts. Our analysis has shown us that we must know the voltage to within $\pm.05$ volts. How many bits of resolution does our system need?

The first thing to do is to express the resolution as a ratio of the minimum value to the maximum range:

$$r_{min} = \frac{V_{min}}{V_{max}}$$
$$= \frac{0.05 \text{ volts}}{10 \text{ volts}}$$
$$= 0.005 \tag{3.3}$$

We can now use Equation 3.2 to find the number of bits. In practice, we would probably try a couple of values of n until we found the right value. A more formal approach, however, would be to solve Equation 3.2 for n:

$$r_{min} = \frac{1}{2^n - 1}$$
$$2^n = \frac{1}{r_{min}} + 1$$
$$n = \log_2 \left(\frac{1}{r_{min}} + 1 \right) \tag{3.4}$$

Plugging in 0.005 for r_{min} into Equation 3.4 yields a value for n of 7.651. Rounding this value up gives a value of eight bits. Therefore, we need to specify *at least* eight bits of resolution for our signal monitor. As a side note, most calculators do not have a $\log_2$ function. The following identity is handy for such situations:

$$\log_b(x) = \frac{\ln(x)}{\ln(b)} \tag{3.5}$$

In this example, we lightly skipped over the method for determining that we needed a resolution of 0.005 volts. Sometimes determining the resolution is straightforward, but sometimes it is not. As a general guide, you can make the following assumptions: Eight bits is adequate for coarse applications. This includes control applications that are not particularly sensitive, and signals that can tolerate a lot of distortion. Eight-bit resolution is adequate for low-grade speech applications, but twelve-bit resolution is much more common. This resolution is generally adequate for most instrumentation and control applications. Twelve-bit resolution produces telephone-quality speech. Sixteen-bit resolution is used for high-accuracy requirements. CD audio is recorded with 16-bit resolution. It turns out that 21 bits is about the maximum practical value for either an ADC or a DAC. Achieving this resolution is expensive, so 21-bit resolution is generally reserved for very demanding applications.

One final word is required on the subject of resolution in terms of the number of bits. The effect of quantizing a signal is to introduce noise. This noise is called, naturally enough, the *quantization error*. The noise can be thought of as the result of representing the smooth and continuous waveform with the stair-step shape of the digitally represented signal.

INSTANT SUMMARY

The overall idea behind digital signal processing is to:

- Acquire the signal.
- Convert it to a sequence of digital numbers.
- Process the numbers as required.
- Transmit or save the data as may be required.
- Convert the processed sequence of numbers back to a signal.

The performance of digital signal processing algorithms is generally specified by frequency response over a normalized frequency range of $-\pi$ to $+\pi$. The actual analog frequencies are scaled over this range by multiplying the digital frequency by the sample period. Accurately representing an analog signal in digital form requires that we convert from the digital domain to the analog domain (or the other way around) with sufficient resolution. In terms of the number of cycles, we must sample at a minimum of *greater than* twice the frequency of the sine wave. The resolution in terms of the amplitude depends upon the application.

Chapter 4

The Math of DSP

In an Instant

- Definitions
- Numerical Concepts
- Complex Numbers
- Causality
- Convolution

- Fourier Series
- Orthogonality
- Quadrature
- Instant Summary

Definitions

So far we have covered many of the "big picture" aspects of DSP system design. However, the heart of DSP is, naturally enough, numbers. More specifically, DSP deals with how numbers are processed. Most texts on DSP either assume that the reader already has a background in numerical theory, or they add an appendix or two to review complex numbers. This is unfortunate, since the key algorithms in DSP are virtually incomprehensible without a strong foundation in the basic numerical concepts.

Since the numerical foundation is so critical, we begin our discussion of the mathematics of DSP with some basic information. This material may be review for many readers. However, we suggest that you at least scan the material presented in this section, as the discussions that follow will be much clearer.

First, let's begin by defining some terms used in this chapter. As you probably remember from beginning calculus, a *function* is a rule that assigns to each element in a set one and only one element in another set. The rule can be specified by a mathematical formula or by tables of associated numbers. A *complex number* is a number of the form a + bj, having a "real" part a and an "imaginary" part bj, with j representing the square root of −1 (although the word "imaginary" doesn't mean that part of the number isn't useful in the real world). A *causal signal* is a signal that has a value of zero for all negative numbered samples. We'll encounter many other important terms in this chapter, but we'll define those as we use them.

FUNCTIONS

In general, applied mathematics is a study of functions. Primarily, we are interested in how the function behaves directly. That is, for any given input, we want to know what the output is. Often, however, we are interested in other properties of a given function. For example, we may want to know how rapidly the function is changing, what the maximum or minimum values are, or how much area the function bounds.

Additionally, it is often handy to have a couple of different ways to express a function. For some applications, one expression may make our work simpler than another.

Polynomials are the workhorse of applied mathematics. The simplest form of the polynomial is the simple linear equation:

$$y = mx + b \tag{4.1}$$

where m and b are constants. For any straight line drawn on an x-y graph, an equation in the form of Equation 4.1 can be found. The constant m defines the slope, and b defines the y-intercept point. Not all functions are straight lines, of course. If the graph of the function has some curvature, then a higher-order function is required. In general, for any function, a polynomial can be found of the form:

$$f(x) = ax^n + \cdots + bx^1 + cx^0 \tag{4.2}$$

which closely approximates the given function, where a, b, and c are constants called the coefficients of $f(x)$.

Insider Info

This polynomial form of a function is particularly handy when it comes to differentiation or integration. Simple arithmetic is normally all that is needed to find the integral or derivative. Furthermore, computing a value of a function when it is expressed as a polynomial is straightforward, particularly for a computer.

If polynomials are so powerful and easy to use, why do we turn to *transcendental* functions such as the sine, cosine, natural logarithm, and so on? There are a number of reasons why transcendental functions are useful to us. One reason is that the transcendental forms are simply more compact. It is much easier to write:

$$y = \sin(x) \tag{4.3}$$

than it is to write the polynomial approximation:

$$f(x) = x - \frac{1}{3!}x^3 + \frac{1}{5!}x^5 - \cdots \tag{4.4}$$

Another reason is that it is often much easier to explore and manipulate relationships between functions if they are expressed in their transcendental form.

For example, one look at Equation 4.3 tells us that $f(x)$ will have the distinctive shape of a sine wave. If we look at Equation 4.4, it's much harder to discern the nature of the function we are working with. It is worth noting that, for many practical applications, we do in fact use the polynomial form of the function and its transcendental form interchangeably. For example, in a spreadsheet or high-level programming language, a function call of the form:

$$y = \sin(x) \tag{4.5}$$

results in y being computed by a polynomial form of the sine function.

Often, polynomial expressions called *series expansions* are used for computing numerical approximations. One of the most common of all series is the Taylor series. The general form of the Taylor series is:

$$f(x) = \sum_{n=0}^{\infty} a_n x^n \tag{4.6}$$

Again, by selecting the values of a_n, it is possible to represent many functions by the Taylor series. In this book we are not particularly interested in determining the values of the coefficients for functions in general, as this topic is well covered in many books on basic calculus. The idea of series expansion is presented here because it plays a key role in an upcoming discussion: the z-transform.

A series may converge to a specific value, or it may diverge. An example of a convergent series is:

$$f(n) = \sum_{n=0}^{\infty} \frac{1}{2^n} \tag{4.7}$$

As n grows larger, the term $1/2^n$, grows smaller. No matter how many terms are evaluated, the value of the series simply moves closer to a final value of 2.

A divergent series is easy to come up with:

$$f(n) = \sum_{n=0}^{\infty} 2^n \tag{4.8}$$

As n approaches infinity, the value of $f(n)$ grows without bound. Thus, this series diverges.

It is worth looking at a practical example of the use of series expansions at this point. One of the most common uses of series is in situations involving growth. The term *growth* can be applied to either biological populations (herds, for example), physical laws (the rate at which a capacitor charges), or finances (compound interest).

Let's take a look at the concept of compound growth. The idea behind it is simple:

- You deposit your money in an account.
- After some set period of time (say, a month), your account is credited with interest.
- During the next period, you earn interest on both the principal *and* the interest from the last period.
- This process continues as described above.

Your money keeps growing at a faster rate, since you are earning interest on the previous interest as long as you leave the money in the account.

Mathematically, we can express this as:

$$f(x) = x + \frac{x}{c} \tag{4.9}$$

where c is the interest rate. If we start out with a dollar, and have an interest rate of 10% per month, we get:

$$f(1) = 1 + \frac{1}{10}$$
$$= 1.10$$

for the first month. For the second month, we would be paid interest on $1.10:

$$f(1.10) = 1.10 + \frac{1.10}{10}$$
$$= 1.21$$

and so on. This type of computation is not difficult with a computer, but it can be a little tedious. It would be nice to have a simple expression that would allow us to compute what the value of our money would be at any given time. With some factoring and manipulation, we can come up with such an expression:

$$f(n) = \left(x + \frac{x}{c}\right)^n \tag{4.10}$$

where n is the number of compounding periods. Using Equation 4.10 we can directly evaluate what our dollar will be worth after two months:

$$f(2) = \left(1 + \frac{1}{10}\right)^2$$
$$= 1.1^2$$
$$= 1.21$$

For many applications, the value of c is proportional to the number of periods. For example, when a capacitor is charging, it will reach half its value in the first time period. During the next time period, it will take on half of the previous value (that is, 1/4), etc. For this type of growth, we can set $c = n$ in Equation 4.10. Assuming a starting value of 1, we get an equation of the following form:

$$f(n) = \left(1 + \frac{1}{n}\right)^n \tag{4.11}$$

Key Concept

Equation 4.11 is a *geometric series*. As n grows larger, $f(n)$ converges to the irrational number approximated by 2.718282. (You can easily verify this with a calculator or spreadsheet.) This number comes up so often in mathematics that is has been given its own name: e. Using e as a base in logarithm calculations greatly simplifies problems involving this type of growth. The natural logarithm (ln) is defined from this value of e:

$$\ln(e) = 1 \tag{4.12}$$

It is worth noting that the function e^x can be rewritten in the form of a series expansion:

$$e^x = 1 + x + \frac{x^2}{2!} + \cdots \frac{x^n}{n!} + \cdots \tag{4.13}$$

The natural logarithm and the base e play an important role in a wide range of mathematical and physical applications. We're primarily interested in them, however, for their role in the use of imaginary numbers. This topic will be explored later in this chapter.

LIMITS

Limits play a key role in many modern mathematical concepts. They are particularly important in studying integrals and derivatives. They are covered here mainly for completeness of this discussion.

The basic mathematical concept of a limit closely parallels what most people think of as a limit in the physical world. A simple example is a conventional signal amplifier. If our input signal is small enough, the output will simply be a scaled version of the input. There is, however, a limit to how large an output signal we can achieve. As the amplitude of the input signal is increased, we will approach this limit. At some point, increasing the amplitude of the input will make no difference on the output signal; we will have reached the limit.

Mathematically, we can express this as:

$$v_{out_{max}} = \lim_{x \to v_{in_{max}}} f(x) \tag{4.14}$$

where $f(x)$ is the output of the amplifier, and $v_{in_{max}}$ is the maximum input voltage that does not cause the amplifier to saturate.

Limits are often evaluated under conditions that make mathematical sense, but do not make intuitive sense to most us. Consider, for example, the function $f(x) = 2 + \frac{1}{x}$ We can find the value of this function as x takes on an infinite value:

$$\lim_{x \to \infty} \left(2 + \frac{1}{x} \right) = 2$$

Key Concept

In practice, what we are saying here is that as x becomes infinitely large, then $1/x$ becomes infinitely small. Intuitively, most people have no problem with dropping a term when it no longer has an effect on the result. It is worth noting, however, that mathematically the limit is not just dropping a noncontributing term; the value of 2 is a mathematically precise solution.

INTEGRATION

Many concepts in DSP have geometrical interpretations. One example is the geometrical interpretation of the process of integration. Figure 4.1 shows how this works. Let's assume that we want to find the area under the curve $f(x)$. We start the process by defining some handy interval—in this case, simply $b - a$. This value is usually defined as Δx. For our example, the interval Δx remains constant between any two points on the x-axis. This is not mandatory, but it does make things easier to handle.

Now, integration is effectively a matter of finding the area under the curve $f(x)$. A good approximation for the area in the region from a to b and under the curve can be found by multiplying $f(a)$ by Δx. Mathematically:

$$\int_a^b f(x)dx \approx f(a)\,\Delta x \tag{4.15}$$

Our approximation will be off by the amount between the top of the rectangle formed by $f(a)\Delta x$ and yet still under the curve $f(x)$. This is shown as a shaded region in Figure 4.1. For the interval from a to b this error is significant. For

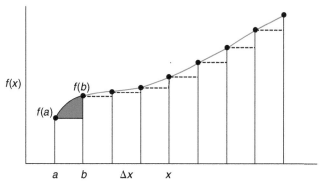

FIGURE 4.1 Geometric interpretation of integration

some of the other regions this error can be seen to be insignificant. The overall area under the curve is the sum of the individual areas:

$$\int f(x)dx \approx \sum f(x)\Delta x \qquad (4.16)$$

It's worthwhile to look at the source of error between the integral and our approximation. If you look closely at Figure 4.1, you can see that the major factor determining the error is the size of Δx. The smaller the value of Δx, the closer the actual value of the integral and our approximation will be. In fact, if the value of Δx is made vanishingly small, then our approximation would be exact. We can do this mathematically by taking the limit of the right-hand side of Equation 4.16 as Δx approaches 0:

$$\int f(x)dx = \lim_{\Delta x \to 0} \sum f(x)\Delta x \qquad (4.17)$$

Notice that Equation 4.17 is in fact the definition of the integral, not an approximation.

There are a number of ways to find the integral of a function. Numerically, a value can be computed using Equation 4.16 or some more sophisticated approximation technique. For symbolic analysis, the integral can be found by using special relationships or, as is more often the case, by tables. For most DSP work, only a few simple integral relationships need to be mastered. Some of the most common integrals are shown in Table 4.1.

OSCILLATORY MOTION

Virtually all key mathematical concepts in DSP can be directly derived from the study of oscillatory motion. In physics, there are a number of examples of oscillatory motion: weights on springs, pendulums, LC circuits, etc. In general, however, the simplest form of oscillatory motion is the wheel. Think of a point on the rim of a wheel. Describe how the point on the wheel moves mathematically

TABLE 4.1 Most frequently used integrals (where c and a are constants and u and v are functions of x).

$$\int du = u + c$$

$$\int u^{-1}du = \int \frac{du}{u}$$
$$= \ln|u| + c$$

$$\int \sec u \tan u \, du = \sec u + c$$

$$\int (du + dv) = \int du + \int dv$$
$$= u + v + c$$

$$\int \sec u \, du = -\cos u + c$$

$$\int \csc u \cot u \, du = -\csc u + c$$

$$\int a\,du = a\int du = au + c$$

$$\int \cos u \, du = \sin u + c$$

$$\int a^u du = \frac{a^u}{\ln a} + c$$

$$\int u^n du = \frac{u^{n+1}}{n+1} + c$$
$$\text{if } n \neq -1$$

$$\int \sec^2 u \, du = \tan u + c$$

$$\int e^u du = e^u + c$$

$$\int \csc^2 u \, du = -\cot u + c$$

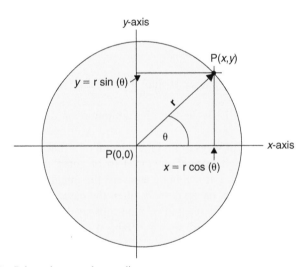

FIGURE 4.2   Polar and rectangular coordinates

and the foundations of DSP are in place. This statement may seem somewhat dramatic, but it is truly amazing how often this simple fact is overlooked.

The natural place to begin describing circular motion is with Cartesian coordinates. Figure 4.2 shows the basic setup. The origin of the coordinate system is, naturally, where the x and y axes intersect. This point is designated as P(0,0). The other interesting point shown in the figure is $P(x,y)$.

The point $P(x,y)$ can be thought of as a fixed point on the rim of a wheel. The axle is located at the point P(0,0). The line from P(0,0) to $P(x,y)$ is a *vector* specified as **r**. We can think of it as the radius of the wheel. (The variable **r** is shown in bold to indicate that it is either a vector or a complex variable.)

The variable **r** is often of interest in DSP, since its length is what defines the *amplitude* of the signal. This will become clearer shortly. When points are specified by their x and y values the notation is called *rectangular*. The point $P(x,y)$ can also be specified as being at the end of a line of length r at an angle of 0. This notation is called *polar* notation.

It is often necessary to convert between polar and rectangular coordinates. The following relationship can be found in any trigonometry book:

$$\text{length of } \mathbf{r} = \sqrt{x^2 + y^2} \tag{4.18}$$

This is also called the magnitude of **r** and is denoted as $|\mathbf{r}|$. The angle θ is obtained from x and y as follows:

$$\theta = \arctan\left(\frac{y}{x}\right) \tag{4.19}$$

Two particularly interesting relationships are:

$$x = |\mathbf{r}| \cos\theta \tag{4.20}$$

and

$$x = |\mathbf{r}| \sin\theta \tag{4.21}$$

The reason these two functions are so important is that they represent the signals we are usually interested in. In order to develop this statement further, it is necessary to realize that the system we have just described is static—in other words, the wheel is not spinning. In DSP, as with most other things, the more interesting situation occurs when the wheels start spinning.

From basic geometry, we know that the circumference of the wheel is simply $2\pi r$. This is important, since it defines the angular distance around the circle. If $\theta = 0$, then the point $P(x,y)$ will have a value of $P(|\mathbf{r}|,0)$. That is, the point will be located on the x-axis at a distance of $|\mathbf{r}|$ from the origin. As 0 increases, the point will move along the dotted line. When $\theta = \pi/2$ the point will be at $P(0,|\mathbf{r}|)$. That is, it will be on the y-axis at a distance $|\mathbf{r}|$ from the origin. The point will continue to march around the circle as θ increases. When θ reaches a value of 2π, the point will have come full circle back to $P(|\mathbf{r}|,0)$.

As the point moves around the circle, the values of x and y will trace out the classic sine and cosine wave patterns. The two patterns are identical, with the exception that the sine lags the cosine by $\pi/2$. This is more often expressed in degrees of phase; the sine is said to lag the cosine wave by 90°.

When we talk about the point moving around the circle, we are really talking about the vector **r** rotating around the origin. This rotating vector is often called a *phasor*. As a matter of convenience, a new variable ω is often defined as:

$$\omega = 2\pi f \tag{4.22}$$

The variable ω represents the *angular frequency*, The variable f is, of course, the frequency. Normally f is expressed in units of *hertz* (Hz), where 1 Hz is equal to 1 cycle per second. (As we have already seen in the last chapter, however, the concept of frequency can take on a somewhat surrealistic aspect when it is used in relation to DSP systems.)

COMPLEX NUMBERS

Now, on to the subject of complex numbers. We have stayed away from this subject until now simply because we did not want to confuse things.

FAQs

How do "imaginary" numbers represent real-world quantities?

Part of the confusion over complex numbers—particularly as they relate to DSP—comes from a lack of understanding over their role in the "real world" (no pun intended). Complex numbers can be thought of as numbers with two parts: the first part is called the *real* part, and the second part is called the *imaginary* part. Naturally, most numbers we deal with in the real world are real numbers: 0, 3.3, 5.0, and 0.33 are all examples. Since complex numbers have two parts, it is possible to represent two related values with one number; *x-y* coordinates, speed and direction, or amplitude and phase can all be expressed directly or indirectly with complex numbers.

Initially, it is easy to think of signals as "real valued." These are what we see when we look at a signal on an oscilloscope, look at a time vs. amplitude plot, or think about things like radio waves. There are no "imaginary" channels on our TVs, after all.

But in practice most of the signals we deal with are actually *complex* signals. For example, when we hear a glass drop we immediately get a sense of where the glass hit the floor. It is tempting to think of the signals hitting our ear as "real valued"—the amplitude of the sound wave reaching our ears as a function of time. This is actually an oversimplification, as the sound wave is really a complex signal. As the glass hits the floor the signal propagates *radially* out from the impact point. Imagine a stone dropped in a pond; its graph would actually be three-dimensional, just as the waves in a pond are three-dimensional. These three-dimensional waves are, in fact, complex waveforms. Not only is the waveform complex, but the signal processing is also complex.

Our ears are on opposite sides of our head to allow us to hear things slightly out of phase. This phase information is perceived by our brains as directional information.

The points we have been discussing, such as P(0,0) and P(x,y), are really complex numbers. That is, they define a point on a two-dimensional plane. We do not generally refer to them this way, however, as a matter of convention. Still, it is useful to remember that fact if things get too confusing when working with complex notation.

> **Insider Info**
>
> *Historically, complex numbers were developed from examining the real number line. If we think of a real number as a point on the line, then the operation of multiplying by (-1) rotates the number $180°$ about the origin on the number line. For example, if the point is 7, then multiplying by (-1) gives us (-7). Multiplying by (-1) again rotates us back to the original value of 7. Thus, the quantity (-1) can be thought of as an operator that causes a $180°$ rotation. The quantity $(-1)^2$ is just one, so it represents a rotation of either $0°$, or equivalently, $360°$.*

This leads us to an interesting question: If $(-1)^2 = 1$, then what is the meaning of $\sqrt{-1}$? There is no truly analytical way of answering the question. One way of looking at it, however, is like this: If 1 represents a rotation of $360°$, and (-1) represents a rotation of $180°$, then $\sqrt{-1}$ must, by analogy, represent a rotation of $90°$. In short, multiplying by $\sqrt{-1}$ rotates a value from the x-axis to the x-axis. Early mathematicians considered this operation a purely imaginary (that is, having no relation to the "real" world) exercise, so it was given the letter i as its symbol. Since i is reserved for current in electronics, most engineers use j as the symbol for $\sqrt{-1}$. This book follows the engineering convention.

> **Key Concept**
>
> In our earlier discussion, we pointed out that a point on the Cartesian coordinates can be expressed as $P(x, y)$. This means, in words, that the point P is located at the intersection of x units on the x-axis, and y units on the y-axis. We can use the j operator to say the same thing:
>
> $$P(x, y) = P(|r| \cos(\theta), |r| \sin(\theta))$$
> $$= x + jy \tag{4.23}$$

Thus, we see that there is nothing magical about complex numbers. They are just another way of expressing a point in the x-y plane. Equation 4.23 is important to remember since most programming languages do not support a native complex number data type, nor do most processors have the capability of dealing directly with complex number data types. Instead, most applications treat a complex variable as two real variables. By convention one is real, the other is imaginary. We will demonstrate this with some examples later.

In studying the idea of complex numbers, mathematicians discovered that raising a number to an imaginary exponent produced a periodic series. The famous mathematician Euler demonstrated that the natural logarithm base, e, raised to an imaginary exponent, was not only periodic, but that the following relationship was true:

$$e^{j\theta} = \cos\theta + j\sin\theta \qquad (4.24)$$

To demonstrate this relationship, we will need to draw on some earlier work. Earlier we pointed out that the sine and cosine functions could be expressed as a series:

$$\sin(x) = x - \frac{x^3}{3!} + \frac{x^5}{5!} - \cdots \qquad (4.25)$$

and

$$\cos(x) = 1 - \frac{x^2}{2!} + \frac{x^4}{4!} - \cdots \qquad (4.26)$$

Now, if we evaluate $e^{j\theta}$ using Equation 4.13 we get:

$$e^{j\theta} = 1 + j\theta - \frac{\theta^2}{2!} - \frac{j\theta^3}{3!} + \frac{\theta^4}{4!} - \frac{j\theta^5}{5!} - \frac{\theta^6}{6!} \cdots \qquad (4.27)$$

Expanding and rearranging Equation 4.27 gives us:

$$e^{-j\theta} = \sum_{m=0}^{\infty} \frac{(-1)^m \, \theta^{2m}}{(2m)!} + j\sum_{m=0}^{\infty} \frac{(-1)^m \, \theta^{2m+1}}{(2m+1)!} \qquad (4.28)$$

Substituting Equation 4.25 and Equation 4.26 into Equation 4.28 gives us Equation 4.24.

Key concept

Euler's relationship is used quite heavily throughout the field of signal processing, primarily because it greatly simplifies analytical calculations. It is much simpler to perform integration and differentiation using the natural logarithm or its base than it is to perform the same operation on the equivalent transcendental functions. Since this book is mainly aimed at practical applications, we will not be making heavy use of analytical operations using e. It is common in the literature, however, to use $e^{j\omega}$ as a shorthand notation for the common $\cos(\omega) + j\sin(\omega)$ expression. This convention will be followed in this book.

Euler's relationship can also be used as another way to express a complex number. For example:

$$P(x, y) = re^{j\theta} \tag{4.29}$$

is equivalent to Equation 4.23.

We have pushed the mechanical analogy about as far as we can, so it is time to briefly review what has been presented and then switch over to an electronic model for our discussion.

- The basic model of a signal is oscillatory motion.
- The simplest conceptualization is a point rotating about the origin.
- The motion of the point can be defined as:

$$P(x, y) = re^{j\omega}$$

where $\omega = 2\pi f$, r is the radius, and f is the frequency of rotation.

- Euler's relationship gives us the following:

$$e^{j\theta} = \cos\theta + j\sin\theta$$
$$e^{-j\theta} = \cos\theta - j\sin\theta$$

The electronic equivalent of the wheel is the LC circuit. An example circuit is shown in Figure 4.3. By convention, the voltage is generally defined as the real value, and the current is defined as the imaginary value. The symbol co is used to represent the resonant frequency and is determined by the value of the components. Assuming the resistance in the circuitry is zero, then:

$$e^{j\omega t} = \cos\omega t + j\sin\omega t \tag{4.30}$$

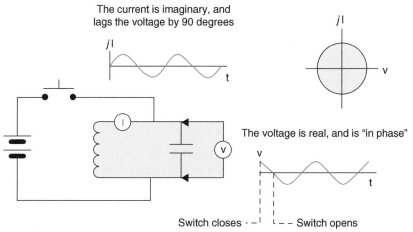

FIGURE 4.3   Ideal LC circuit showing voltage and current relationships

describes the amplitude and the phase of the voltage and the current. In practice, we would add in a scale factor to define the value of the maximum voltage and the maximum current. Notice that, as in the case of the point rotating about the origin, the voltage is 90° out of phase with the current.

What if the resistance is *not* equal to zero? Then the amplitude decreases as a function of time. From any good book on circuit analysis, we can find that the decay of the amplitude is an exponential function of time: $e^{-\alpha t}$. This decay applies to both the current and the voltage. If we add in our scale factor A, we get the following equation:

$$f(t) = A e^{-\alpha t} e^{j\omega t} \tag{4.31}$$

which, from our log identities, gives us:

$$f(t) = A e^{(-\alpha + j\omega)t} \tag{4.32}$$

Generally, the exponential term is expressed as a single complex variable, s:

$$s = -\alpha + j\omega \tag{4.33}$$

The symbol s is familiar to engineers as the independent variable in the Laplace transform. (Transforms will be covered in the next chapter.)

How It Works

In order to illustrate some of the basic principles of working with discrete number sequences, we will begin with a simple example. Referring back to Figure 4.1, let's assume that our task is to use a DSP system to generate a sine wave of 1 Hz. We will also assume that our DAC has a resolution of 12 bits, and an output range of −5 volts to +5 volts.

This task would be difficult to do with conventional electronic circuits. Producing a sine wave generally requires an LC circuit or a special type of RC oscillator known as a Twin-T. In either case, finding a combination of values that work well and are stable at 1 Hz is difficult.

On the other hand, designing a low-frequency oscillator like this with DSP is quite straightforward. We'll take a somewhat convoluted path, however, so we can illustrate some important concepts along the way.

First, let's look at the basic function we are trying to produce:

$$f(t) = \sin(\omega t + \theta) \tag{4.34}$$

where, for this example, $\omega = 2\pi f$, $f = 1$, and $\theta = 0$.

From a purely mathematical perspective. Equation 4.34 is seemingly simple. There are some interesting implications in this simple-looking expression, however. As Rorabaugh[1] points out, the notation $f(t)$ is used to mean different

---

[1] Digital *Filter Designers Handbook,* page 36 (see References).

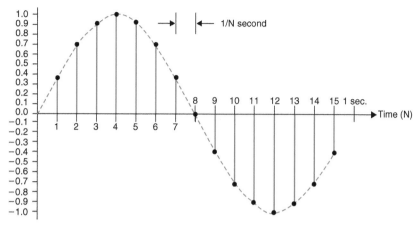

FIGURE 4.4   Sample points on a sine wave

things by various authors. It may mean the entire function expressed over all values of t, or it may mean the value of f evaluated at some point t.

Another interesting concept is the idea that $f(t)$ is continuous. In practice, we know that no physical quantity is truly infinitely divisible. At some point quantum physics—if no other physical law—will define discretely quantized values. Mathematically, however, $f(t)$ is assumed to be *continuous, and therefore infinitely divisible.* That is, for any $f(t)$ and any/ $f(t + \Delta)$ there is some value equal to $f(t + \Delta/2)$. This leads to the rather interesting situation that between any two *finite* points on a line there are an *infinite* number of points.[2]

The object is to use a digital computer to produce an electrical output representing Equation 4.34. Clearly, we cannot compute an infinite number of points, as this would take an infinite length of time. We must choose some reasonable number of points to compute. What is a "reasonable number of points"? The answer depends on the system we are using and on how close an approximation we are willing to accept. In practice we will need something like 5 to 50 points per cycle. Figure 4.4 shows an example of how 16 points can be used to approximate the shape of a sine wave. Each point is called one *sample* of the sine function (N = 15).

Notice that time starts at $t = 0$ and proceeds through $t = {}^{15}\!/_N$. In other words, there are 16 points, each evaluated at $\frac{1}{16}$ -second intervals. This interval between samples is called (naturally enough) the sample period. The sample period is usually given the symbol T. Notice that the next cycle starts at $t = 0$ *of the second cycle*, so there is no point at the 1-second index mark. In order to incorporate T in an equation we must define a new term: the *digital frequency,* In our discussion of the basic trigonometry of a rotating point, we

---

[2] See pages 152–157 of The *Mathematical Experience* for a good discussion of this.

defined the angular frequency, ω, as being equal to $2\pi f$. The digital frequency λ is defined as the analog frequency times the period T:

$$\lambda = \omega T$$
$$= \frac{\omega}{N} \tag{4.35}$$

The convention of using λ as the digital frequency is not universal; giving the digital frequency its own symbol is useful as a means of emphasizing the difference between the digital and the analog frequencies, but is also a little confusing. In this text we denote the digital frequency as ωT. The justification for defining the digital frequency in this way will be made clear shortly.

The variable t is continuous, and therefore is not of much use to us in the computations. To actually compute a sequence of discrete values we have to define a new variable, n, as the index of the points. The following substitution can then be made:

$$t = nT, n = 0...N - 1 \tag{4.36}$$

Equation 4.35 and Equation 4.36 can be used to convert Equation 4.34 from continuous form to a discrete form. Since our frequency is 1 Hz, and there is no phase shift, the equation for generating the discrete values of the sine wave is then:

$$
\begin{aligned}
f(t) &= \sin(2\pi f t + \theta)\big|_n \\
&= \sin(2\pi(1)nT + 0), n = 0...N - 1 \\
&= \sin(2\pi nT), n = 0...N - 1
\end{aligned}
\tag{4.37}
$$

Remember that T is defined as $1/N$. Therefore, Equation 4.37 is just evaluating the sine function at 0 to $^{N-1}/N$ discrete points. The need to include T in Equation 4.37 is the reason that the digital frequency was defined in Equation 4.35.

For a signal this slow, we could probably compute the value of each point in *real time*. That is, we could compute the values as we need them. In practice, however, it is far more efficient to compute all of the values ahead of time and then save them in memory. The first loop of the listing in Figure 4.5 is an example of a C program to do just this.

The first loop in Figure 4.5 generates the *floating point* values of the sine wave. The DAC, however, requires binary integer values to operate properly, so it is necessary to convert the values in k to properly formatted integers. Doing this requires that we know the binary format that the DAC uses, as there are a number of different types. For this example, we will assume that a 0 input to the DAC causes the DAC to assume its most negative (-5 V) value. A hexadecimal value of 0 × FFF (that is, all ones) will cause the most positive output ($+5$ V).

The floating point values in $k[\ ]$ have a range of -1.0 to $+1.0$. The trick then is to convert these values so that -1.0 maps to 0 × 000 and $+1.0$ maps to

```
#include <stdio.h>
#include <math.h>

/* Define the number of samples. */
#define N 16

void main()
{

    unsigned int DAC_values[N]; /* Values used by the DAC. */

    double k[N]; /* Array to hold the floating point values. */
    double pi; /* Value of pi. */

    /* Declare an index variable. */
    unsigned int n;

            pi = atan(1) * 4; /* Compute the value of pi.*/

            for (n=0; n<N; n++)
                {
                k[n] = sin(2 * pi * ((float)n/(float)N));
                printf("%1.2f\n",k[n]);
                }
            for (n=0; n<n; n++)
                {
                DAC_values[n] = ((k[n] / 2.0) + 0.5) * OxFFF;
                printf("%3X\n",DAC_values[n]);
                }
//   The following code is system dependent, so we have
//   provided pseudo-code to illustrate the types of things
//   that need to be done. The functions wait_seconds() and
//   Output_to_DAC() are user defined.
//     while (1) /* Set up an infinite loop. */
//     {
//       for(n=0;  n<N; n++)
//        {
//         wait_seconds (1/ (float) N); /* Wait 1/N seconds.*/
//         Output_to_DAC(DAC_value6[n]); /* Output each value. */
//        }
//      }
//
}
```

FIGURE 4.5   C listing for generating a sine wave

$0 \times$ FFF. We can do this by dividing all of the values in k by 2, and then adding 0.5. This scales the values in k from 0.0 to 1.0. Then, we can multiply the values in k by $0 \times$ FFF. The result is a series of binary integers that represent equivalent values of the waveform. This operation is shown in the second loop of Figure 4.5.

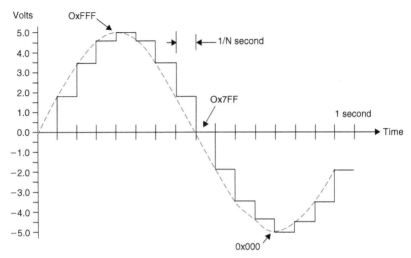

FIGURE 4.6   DAC output for a sine wave

The final step is to periodically (every $T = 1/N$ seconds) output the indexed value of $k[\ ]$. This step is highly system dependent, so it is not practical to present real code to perform the output function. At the bottom of Figure 4.5 is pseudocode that shows a typical sequence, however.

The result is shown in Figure 4.6. The stair-step shape is the output of the DAC. The dashed line is the ideal sine wave. After passing through the smoothing filter, the actual waveform will approximate the ideal.

This example is straightforward, but it does illustrate some very important concepts. One of these is, as we noted earlier, the concept *of digital frequency* vs. *analog frequency.* Previously we just defined the digital frequency as ωT, where T is equal to $1/N$ seconds, and N is the number of samples per seconds. In many practical applications, however, there is really no need to keep the relationship $T = 1/N$. For example, we can just assume that $T = 1$. Then, all we really care about is the ratio $^n/N$; the value of T simply becomes a scaling factor. Another example will help illustrate the point.

In our previous example, we built a function generator, using digital techniques, to output a sine wave of 1 Hz. In that example, the digital and the analog frequency were the same thing. Now, let's consider how to modify the output frequency of the function generator. There are actually two ways to accomplish this.

Let's assume we want to double the output frequency, from 1 Hz to 2 Hz. The first way to do this would be to *decrease the time* we wait to output the next sample to the DAC. For example, instead of waiting $1/N$ seconds to output the new value to the DAC, we could wait only $\frac{1}{2}N$ seconds to output the value. This would *double* the number of points that are output each second. Or, equivalently, we could think of this as outputting one cycle of the waveform in 0.5 seconds.

The important thing to notice here is that we have not reevaluated Equation 4.37. We have changed the value of T but, as long as we understand what the implications are, there is no need to recompute the values of $f[n]$. The actual frequency output, interestingly enough, has *nothing* to do with the values computed. The actual (analog) frequency will match the digital (computed) frequency only when the output interval between points is equal to $1/N$ seconds. In this sense we see that digital frequency is computationally independent of the analog frequency.

This may seem a bit obtuse and esoteric, but it is of practical importance. Many DSP applications do not require real-time evaluation. For example, in seismic analysis the data is recorded first, and then processed. Processing a sample generally takes much longer than the time over which the signal was recorded. A 10-second record, for example, may take hours or days of computation time to process. In such situations, the value of T is critical only in scaling the final results. What counts computationally is the value N.

Key Concept

In many DSP applications, the number of samples per some "unit period" determines how the signal is handled. Once processed, the signal is mapped back into real time by a scale factor T. T may or may not be directly related to $1/N$ seconds.

What is the second way to change the output frequency? We could leave the output interval at $1/N$ seconds, and change the value of f in Equation 4.37. If we let $f = 2$, then Equation 4.37 becomes:

$$
\begin{aligned}
f(t) &= \sin(2\pi f t + \theta)\big|_n \\
&= \sin(2\pi(2)nT + 0), n = 0\ldots N - 1 \\
&= \sin(4\pi nT), n = 0\ldots N - 1 \\
&= \sin\left(\frac{4\pi n}{N}\right), n = 0\ldots N - 1
\end{aligned}
\tag{4.38}
$$

Notice that there will now be *two* cycles in 16 points. Each cycle of the sine wave will only have 8 points, as shown in Figure 4.7.

This approach has the advantage that no adjustments have to be made to the output interval timing routines. On the other hand, the quality of the output waveform will vary as a function of frequency. This is because the number of points per cycle varies as a function of frequency. A practical DSP system must balance, and sometimes adjust in real time, the tradeoffs between the number of points used per second and the time interval between each point.

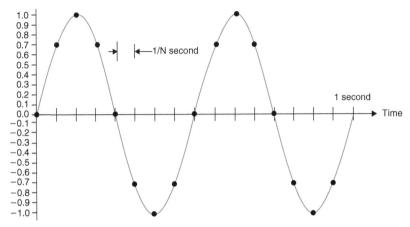

FIGURE 4.7   Two cycles of a sine wave

EXAMPLE APPLICATIONS

At this point let's take a look at where we have been and where we are going. So far, we've been concerned with the mechanics of getting a signal into and out of our DSP system, and with reviewing some general math principles we will use later on. We have seen that we can sample a waveform, optionally store it, and then send it back out to the world. This is, in and of itself, a very useful ability. However, it represents only a small fraction of the things we can do with a DSP system.

Understanding how a DSP system is designed and used basically requires two types of knowledge. The first is an understanding of the applications that lend themselves best to DSP. The second type is an understanding of the tools necessary to design the system to accommodate these applications.

With this in mind, let's now turn our attention to the subject of filtering, beginning with a simple filter that is easily understood intuitively. We will then move on to developing the tools and techniques that will allow us to create more sophisticated, higher-performance filters of professional quality.

FILTERS

One of the most common DSP operations is filtering. As with analog filters, DSP filters can provide low-pass, bandpass, and high-pass filtering. (Specialized functions, such as notch filters, are also possible, though we will not be covering them in this book.) The basic idea behind filtering in general is this: An input signal, generally a function of time, is input to a *transfer function*. Normally, the transfer function is a differential equation expressed as a function of frequency. The output of the transfer function is some *subset* of the input signal.

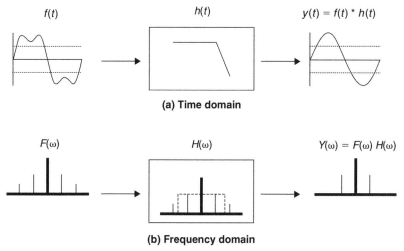

FIGURE 4.8   The basic low-pass filter

A block diagram of a low-pass filter is shown in Figure 4.8. In the figure, the input signal is a sum of two sine waves: one of them at a fundamental frequency, the other at the third harmonic. After passing through the transfer function $H(\omega)$ only the fundamental frequency remains; the first harmonic has been blocked. The top portion of Figure 4.8 depicts the low-pass filter as a function of time. The bottom portion of Figure 4.8 shows the filter as a function of frequency. We will be revisiting these concepts in greater detail in later chapters.

> **Key Concept**
>
> In the world of analog electronics, the transfer function $H(\omega)$ is realized by arranging a combination of resistors, capacitors, inductors, and possibly operational amplifiers. In DSP applications, a computer is substituted for the resistors, capacitors, and inductors. The computer then computes the output using the input and $H(\omega)$.

The question for the DSP applications developer then becomes: How do we define $H(\omega)$ to give us the desired transfer function? This chapter shows, in an intuitive way, how simple digital filters operate. After that, several key concepts are introduced that lay the groundwork for developing more sophisticated filters. In the next chapters, we will see how to apply these tools to develop some practical working filters.

Example 1

First, let's examine a simple application. Consider, for example, that much of the most interesting music of the twentieth century is stored on phonograph

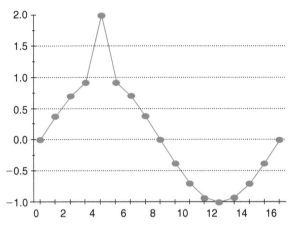

FIGURE 4.9   A noise "pop" on a sine wave

records. These records store their data using variations in the groove running from the outside of the record to its center. Over time, peaks in the groove can break off, or dents can be forced in the walls of the groove. When the phonograph needle hits one of these obstructions, the result is a "pop" in the music being played, as shown graphically in Figure 4.9. A pop is shown riding on an otherwise clean sine wave.

As these records are converted to digital form, it is natural to look for ways to eliminate these pops, thus restoring the more natural sound of the recording. One obvious solution is to manually adjust the spike down to a level where it is consistent with the rest of the signal. This could be done with a waveform editor or, in this simple case, even with a spreadsheet program.

Insider Info

Actually, manually editing the waveform is a good approach since it makes use of the best signal processor in the world: the human brain. For critical passages, it is fairly common for a person to manually edit the waveform. However, this approach is quite labor intensive. CDs, for example, are sampled at 44 kHz, and manually searching 44,000 points for each second of music rapidly becomes prohibitive. It s reasonable to find a more automated approach.

One simple approach is to average the value on either side of the spike with the value of the spike. This would not eliminate the spike, but it certainly would minimize it. We can do this using a simple algorithm:

$$g(n) = \frac{f(n-1) + f(n) + f(n+1)}{3} \qquad (4.39)$$

TABLE 4.2 Result of applying averaging routine to signal in Figure 4.9

n	f(n)	$\dfrac{f(n-1)+f(n)+f(n+1)}{3}$
−1	0.000	0.000
0	0.000	0.128
1	0.383	0.363
2	0.707	0.671
3	0.924	1.210
4	2.000	1.283
5	0.924	1.210
6	0.707	0.671
7	0.383	0.363
8	0.000	0.000
9	−0.383	−0.363
10	−0.707	−0.671
11	−0.924	−0.877
12	−1.000	−0.949
13	−0.924	−0.877
14	−0.707	−0.671
15	−0.383	−0.363
16	0.000	−0.128
17	0.000	0.000

Table 4.2 shows what happens when we apply this averaging routine to the signal in Figure 4.9. Notice that we have applied the averager across the entire signal from $n = -1$ to $n = 17$. This has the effect of moving the center point along the waveform. Therefore, this type of filter is known as a *moving average* filter. Notice that the table actually starts before the first sample—that is, we start evaluating $g(n)$ for $n = -1$. This might seem a little strange, but it makes sense when you consider that one of the terms in $g(n)$ is $f(n + 1)$. By starting at $n = -1$, we begin evaluating the signal at $f(0)$. For the first output value that we compute, $n = -1$, we have defined $f(-2)$ and $f(-1)$ to be zero. In a similar fashion, the value of $f(n + 1)$ is defined to be zero when $n = 16$ and $n = 17$.

The averaged values closely track the original values except at $n = 4$. For $n = 4$ the average value is much smaller than the input value. It is, in fact,

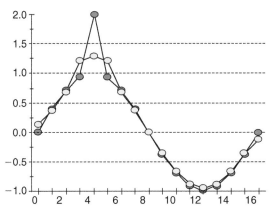

FIGURE 4.10   Effects of a moving average filter

much closer t o where we want it. This routine does a fairly good job of mini-
mizing the pops in a recording. Figure 4.10 is a graph of the original function
and the output of our averaging routine.

Let's look more closely at how and why this routine works. Most of the
changes in values from one point to the next point in the original signal are relatively small. Therefore, for most points, the average value of the three points
is relatively close to the center value. At $n = 4$ in the original signal, however,
the value makes a large (or, equivalently, *rapid*) change. The moving average
routine prevents this rapid change from propagating through.

> **Key Concept**
>
> The action of the averager has little effect on slowly changing signals and a much
> larger effect on rapidly changing signals. This is equivalent to saying that low-fre-
> quency signals suffer little attenuation, while high-frequency signals are strongly
> attenuated. That is, of course, the definition of a *low-pass filter*.

While it is clear that Equation 4.39 represents a low-pass filter, it is *not* clear
exactly what the frequency response of the filter is. One conceptually simple
way to find the frequency response of this filter is to measure the response for
a variety of sinusoidal inputs. For example, let's divide the frequencies between
0 and π into six frequencies. Next, feed into the filter cosine waveforms at
these frequencies and measure the peak output. We picked a *cosine* waveform
because it gives us a value of 1 for an input of 0 Hz, keeping the response consistent with a low-pass filter. With this information, we can then create a table
of the frequency response, as shown in Table 4.3. From this table we can graph
the frequency response of our low-pass filter; the graph is shown in Figure 4.11.

So far our development of the low-pass filter, and its response, has been
very empirical This is often how it is done in the real world. For example, the

TABLE 4.3 **Frequency response**

Frequency (cosine wave input)	Response (peak amplitude)
0.000	1.000
$\pi/5$	0.873
$2\pi/5$	0.539
$3\pi/5$	0.373
$4\pi/5$	0.167
π	0.000

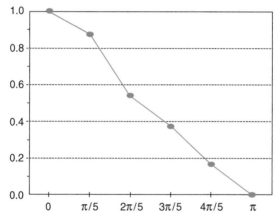

FIGURE 4.11   Frequency response of a simple filter

financial community often makes use of moving averages to filter out the day-to-day variations in stock prices, commodity prices, etc. This filter allows the stock analysts to see the underlying trend of the price, without having the trend line distorted by transient perturbations.

Insider Info

A sine function with an input of 0 Hz produces an output of 0. A cosine function with an input of 0 Hz produces an output of 1. Had we used a sine wave, the 0 Hz input value would have produced an output value of 0. This is mathematically acceptable, but it would not be consistent with generating test data for a low-pass filter. In this respect, a sine wave of 0 Hz is a bit anomalous. This situation of switching between a sine and a cosine wave is a fairly common trick in the literature.

On the other hand, this empirical approach can be difficult to manage for more sophisticated filters. As can be seen from Figure 4.11, the moving average filter is not very "crisp." It gradually attenuates the signal as the frequency increases. Often, we are more interested in a "brick wall" filter, which is a filter that does not affect the signal at all up to a cutoff frequency, then reduces any higher frequency components to zero above the cutoff.

Shortly we will look at more formal ways of developing and evaluating filters. But first let's explore these intuitive filters a little more.

Example 2

Let's revisit Figure 4.9. On our last pass the signal was the sine wave and the noise was the spike. It could just as easily have been the other way around, however. For example, one problem that constantly plagues engineers is the presence of the 60-Hz "hum" created by the ubiquitous AC power wiring. This problem generally manifests itself as a sine wave superimposed on top of the signal of interest. A typical example is a system that monitors photons. When a photon strikes a detector, it produces a small electrical pulse. The result of such a pulse on top of the 60-Hz hum would look like Figure 4.9.

How can we eliminate the 60-Hz hum and leave the signal relatively intact? Obviously, our moving average filter will not do the job in this case. It does, however, suggest a solution. If we took the average of the points, and then *subtracted* this average value from the center value, we get the desired result. Algorithmically:

$$g(n) = f(n) - \frac{f(n-1) + f(n) + f(n+1)}{3} \qquad (4.40)$$

Table 4.4 shows the results of applying Equation 4.40 to the data shown in Figure 4.9. The graphical result is shown in Figure 4.12. Notice that the sine wave is essentially eliminated, leaving only the spike. Just as the moving average filter represented a low-pass filter, this differential filter represents a *high-pass* filter; the low-frequency sine wave is heavily attenuated, the high-frequency spike is only moderately attenuated.

These two examples illustrate in an intuitive way how digital filters work. In practice, most common digital filters are simply more sophisticated versions of these simple filters. A bandpass filter, for example, can be achieved by combining a low-pass filter and a high-pass filter.

CAUSALITY

Key Concept

Causality refers to whether a filter can be implemented in *real time*. For example, in a *causal* DSP system that changes an input signal into an output signal, the value at sample number 6 of the input signal can affect only sample number 6 or higher in the output signal.

TABLE 4.4 Results of applying Eq. 4.40 to data in Figure 4.9

n	f(n)	f(n) − f(n − 1) + f(n) + f(n + 1)
−1	0.000	0.000
0	0.000	−0.128
1	0.383	0.019
2	0.707	0.036
3	0.924	−0.286
4	2.000	0.717
5	0.924	−0.286
6	0.707	0.036
7	0.383	0.019
8	0.000	0.000
9	−0.383	−0.019
10	−0.707	−0.036
11	−0.924	−0.047
12	−1.000	−0.051
13	−0.924	−0.047
14	−0.707	−0.036
15	−0.383	−0.019
16	0.000	0.128
17	0.000	0.000
18	0.000	0.000

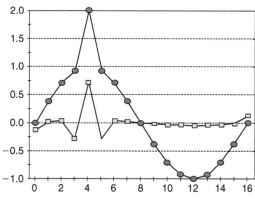

FIGURE 4.12   Effects of a difference filter

Looking back at our moving average filter, notice that for any given sample n, we used both $n - 1$ and $n + 1$ sample points as well. If we think about n as being the current sample (that is, the one coming in immediately), we obviously have a problem. Getting the $n + 1$ sample means that we must know the *future* value of f.

In our recording example, this was not a problem. Since the data is recorded, we can find values for points that appear, with respect to n, to be both in the future $(n + 1)$ and in the past $(n - 1)$. For a real-time application, however, this is not an option; we are constrained to using only the current and past values of f. Filters that require only current and past values of a signal are called *causal filters*. Filters, such as our moving average filter, that require future values are called *noncausal* filters. As a matter of perspective, all real-world analog filters are causal. This is another example of the advantage of DSP: it allows us to build filters that could not be realized in any other way.

Notice that we can make our filter causal by simply shifting the index back by one:

$$y(n - 1) = \frac{f(n + 1) + f(n) + f(n - 1)}{3} \tag{4.41}$$

which is equivalent to:

$$y(n) = \frac{f(n) + f(n - 1) + f(n - 2)}{3} \tag{4.42}$$

Equation 4.42 will not work quite as well as the noncausal version, since it is not symmetrical about the sample point. It will work nearly as well, however. In fact, the difference may be virtually undetectable in many applications. More important for our discussion is the fact that it does not significantly change our *conceptualization* of how the moving average filter works.

CONVOLUTION

Key Concept

Convolution, in its simplest terms, is the process of feeding one function into (or as it is sometimes called, *through*) another function. Conceptually, for example, a filter can be thought of as a function. When we feed some function (such as the one in Figure 4.9) through our moving average filter, we are *convolving* the input function with the moving average filter. The asterisk (*) is normally used to denote convolution:

$$y[n] = f[n]*h[n] \tag{4.43}$$

where $h[n]$ are the coefficients of our filter, and $f[n]$ is the input function. In our moving average filter $h[n]$ had three coefficients and they were all equal to 1/3.

Convolution is sufficiently important to DSP that it is worth developing the subject in detail. In the following examples, the notation will be somewhat simplified. Instead of using $f[n]$, we will use the simpler f_n. The meaning is identical.

In review then, our moving average filter can be expressed as follows:

$$y[n] = \frac{f[n+1] + f[n] + f[n-1]}{3} \tag{4.44}$$

Distributing the (1/3) gives us:

$$y[n] = \frac{1}{3}f[n+1] + \frac{1}{3}f[n] + \frac{1}{3}f[n-1] \tag{4.45}$$

To make the expression more general, we replace the constants with the function h:

$$y[n] = h_0 f[n+1] + h_1 f[n] + h_2 f[n-1] \tag{4.46}$$

Converting to our simpler notation yields:

$$y_n = h_0 f_{n+1} + h_1 f_n + h_2 f_{n-1} \tag{4.47}$$

It is worthwhile to study the actual computation sequence that goes on in the filter. Let's take the first four samples of f: f_0, f_1, f_2 and f_3.

We start out at time $n = -1$. The first computation is then:

$$y_{-1} = h_0 f_0 + h_1 f_{-1} + h_2 f_{-2} \tag{4.48}$$

Immediately, a problem crops up. We require values of f with a negative index. In other words, we need values *before* our first sample. We can get around this problem by simply defining f to be 0 at any point where it is not explicitly defined. Thus, for $n = -1$ we obtain:

$$y_{-1} = h_0 f_0 \tag{4.49}$$

This notation is still a little awkward, since the y_{-1} implies that our first output occurs at some time prior to the $n = 0$ point. This is just a manifestation of our noncausal implementation. It really is our *first* output.

In a similar fashion, we can get the next output for $n = 0$:

$$\begin{aligned} y_0 &= h_0 f_1 + h_1 f_0 + h_2 f_{-1} \\ &= h_0 f_1 + h_1 f_0 \end{aligned} \tag{4.50}$$

Proceeding along these lines, we obtain the results shown in Table 4.5. Notice the symmetry and pattern of the terms in the table. We have been careful to line up the terms in the equations to emphasize this point. With a little contemplation, we can derive a very compact expression for producing the terms in Table 4.5:

$$y[n] = \sum_{k=-\infty}^{\infty} h[k]f[n-k] \tag{4.51}$$

TABLE 4.5 Results of convolution example

$$y[-1] = h_0 f_0$$
$$y[0] = h_0 f_1 + h_1 f_0$$
$$y[1] = h_0 f_2 + h_1 f_1 + h_2 f_0$$
$$y[2] = \quad\quad\quad h_1 f_2 + h_2 f_1 + h_3 f_0$$
$$y[3] = \quad\quad\quad\quad\quad\quad h_2 f_2 + h_3 f_1$$
$$y[4] = \quad\quad\quad\quad\quad\quad\quad\quad\quad h_3 f_2$$

One caveat: Don't try to apply Equation 4.51 too literally to produce Table 4.5, as the $n = -1$ term will throw you off. If you start with $n = 0$, however, you will get the same *terms* shown in Table 4.5. More formally, we can say that Equation 4.51 is valid for all non-negative index values of y.

Equation 4.51 is called the *convolution sum*, and we can use it directly to implement filters. We simply plug in the coefficients for h, and then feed in the values for the input function f. Obviously, finding the coefficients for h is of key interest. So far we have only been able to come up with the simple moving average filter:

$$h[n] = \frac{1}{N}, n = 0, 1, 2 \ldots N - 1 \tag{4.52}$$

Increasing N gives more terms to average, and therefore a lower frequency response. Fewer terms give fewer terms to average, and therefore a higher frequency response. As we saw, we can empirically determine the curve for the frequency response, but we cannot really do much to control the *shape* of the curve.

It would be much more useful if we could simply draw the frequency response we wanted, and then convert that frequency response to the coefficients for h. That is exactly what we will do, but first we must develop a few more tools.

THE FOURIER SERIES

The Fourier series plays an important theoretical role in many areas of DSP. However, it generally does not play much of a practical role in actual DSP system design. For this reason, we will spend most of this section discussing the insights to be gained from the Fourier series; we will not devote a great deal of time to the mathematical manipulations commonly found in more academic texts.

Figure 4.13 shows an example of how the Fourier series can be used to generate a square wave. The square wave can be approximated by the expression:

$$f(t) = \sin \omega t + \frac{1}{n}\sin(n\omega t), n = 1, 3, 5, 7, \ldots, \infty \qquad (4.53)$$

The first term on the right side of Equation 4.53 is called the fundamental frequency. Each value of n is a *harmonic* of the fundamental frequency.

Looking at Figure 4.13, we can see that after only two terms the waveform begins to take on the shape of a square wave. Adding in the third harmonic produces a closer approximation to a square wave. If we keep adding in harmonics, we continue to obtain a waveform that looks more and more like a square wave. Interestingly enough, even if we added an infinity of odd harmonics we would not get a perfect waveform. There would always be a small amount of "ringing" at the edges. This is called the *Gibbs phenomena.*

There are some very interesting implications to all of this. The first is the fact that the bandwidth of a signal is a function of the *shape* of a waveform. For example, we could transmit a 1-kHz sine wave over a channel having a bandwidth of 1 kHz, but if we wanted to transmit a 1-kHz *square* wave we would have a problem.

Equation 4.53 tells us that we need infinite bandwidth to transmit a square wave! And, indeed, to transmit a *perfect* square wave would require infinite bandwidth. However, a perfect square wave is *discontinuous;* the change from the low state to the high state occurs in zero time. *Any* physical system will require *some* time to change state. Therefore, any attempt to transmit a square wave must involve a compromise.

In practice, 10 to 15 times the fundamental frequency provides enough bandwidth to transmit a high-quality square wave. Thus, to transmit our 1-kHz square wave would require something like a 10-kHz bandwidth channel.

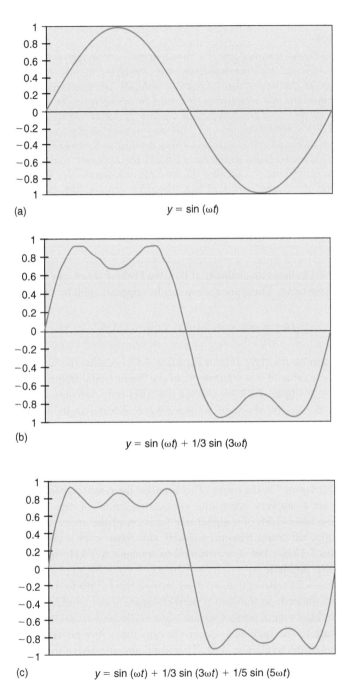

FIGURE 4.13   Creating a square wave from a series of sine waves

A wider channel would give a sharper signal, while a narrower channel would give a more rounded square wave.

These observations lead to some interesting correlations. The higher the frequency that a system can handle, the faster it can change value. Naturally, the converse is true: The faster a system can respond, the higher the frequency it can handle.

This information also gives us the tools to complete the development of the Nyquist theorem.

The Nyquist Theorem Completed

Earlier we demonstrated that we needed at least two nonzero points to reproduce a sine wave. This is a necessary but not sufficient condition. For any two (or more) nonzero points that lie on the curve of a sine wave, there are an infinite number of harmonics of the sine wave that will also fit the same points. We eliminated the harmonic problem by requiring that all of our samples be restricted to one cycle of the sine wave. We will revisit this limitation in a minute, but first let's look closer at our work on the Nyquist theorem up to this point.

The big limitation on our development of the Nyquist theorem so far has been the requirement that we only deal with sine waves.

By taking into account the Fourier series we can remove this limitation. The Fourier series tells us that, for any practical waveform, we can think of it as the sum of a number of sine waves. All we need to concern ourselves with is handling the highest frequency *present* in our signal. This allows us to state the Nyquist theorem in the form normally seen in the literature.

> **Key Concept**
>
> **To accurately reproduce a signal, we must sample at a rate *greater than twice* the frequency of the highest frequency component *present* in the signal.**

The bold emphasis is to highlight two areas that are often misinterpreted. It is often stated that it is necessary to sample at *twice* the highest frequency of interest. As we saw earlier, sampling at twice the frequency only guarantees that we will get two points over one cycle. If these two points occur at the zero crossing, it would be impossible to fit a curve to the two points.

Another common mistake is to assume that it is sufficient to sample a signal at twice the frequency *of interest*. It is not the frequency of interest, but rather the frequency *present* that is important. If there are signal components higher in frequency than the Nyquist frequency, they will be aliased into the frequency below the Nyquist frequency and cause distortion of the sampled signal.

Let's demonstrate with an example. Let's say that we are interested in building a DSP system that can record voices at telephone-quality levels. Generally, telephone-quality speech can be assumed to have a bandwidth of 5 kHz. Even though the human hearing range is generally defined as 20 Hz to 20 kHz, most speech information is contained in the spectrum below 5 kHz.

The limiting factor on an analog voice input is generally the microphone. These typically handle frequencies up to 20 or 30 kHz, though the cheaper mikes will start rolling off in amplitude around 10 kHz or so. Thus, there will be frequency components present that are well above our upper frequency of interest. An anti-aliasing filter is needed to eliminate these components.

If we assume that we want to sample our signal at five times the highest frequency of interest, then our sampling rate would be 25 kHz. Strictly speaking, this would dictate a Nyquist frequency of 12.5 kHz. However, since we are not interested in frequencies this high, it makes sense to set the cutoff of the anti-aliasing filter at around 6 kHz or so. This gives us some headroom above our design requirement of 5 kHz, but is low enough that we will be *oversampling* the signal by a factor greater or equal to 12.5 kHz/6 kHz. This oversampling allows us to relax the performance specifications on the analog parts of the system, thus making our system more robust and easier to build.

Setting the cutoff of the anti-aliasing filter well below the Nyquist frequency has another significant advantage: it allows us to specify a simpler filter with a slower roll-off. Such a filter is cheaper and introduces much less phase distortion.

ORTHOGONALITY

The term *orthogonality* derives from the study of vectors. Most likely you have run across the term in basic trigonometry or calculus. By definition, two vectors in a plane are orthogonal when they are at a 90° angle to each other. When this is the case, the dot product of two vectors is equal to zero:

$$\xrightarrow[v_2]{} \bullet \uparrow v_2 = 0$$

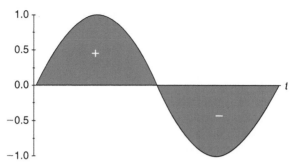

FIGURE 4.14 The average area under a sine wave is zero

The main point here is that the idea of multiplying two things together and getting a result of zero has been generalized in mathematics under the term *orthogonality.*

We will get back to this shortly, but let's look at another case where an interesting function has a zero value: the average value of a sine wave. Figure 4.14 shows one cycle of a sine wave. We have shaded in the area under the curve for the positive cycle and the area above the curve for the negative cycle. Notice that the area for the negative portion of the waveform is labeled with a negative symbol. A "negative area" is a hard concept to imagine, but be reassured that we are simply talking about an area that has a negative sign in front of it. If we add the two areas together we will, naturally, get a value of zero. This may seem too obvious to bother pointing out, but it is just the first step.

> **Insider Info**
>
> *As an interesting side note, this fact was used in the early days of electricity to "prove" that AC voltages were of no practical use. Since they averaged to zero, so the analysis went, they could not do work!*

The process of integration can be viewed as finding the area under a curve. Therefore, you can write this idea mathematically as follows, for any integer value of k:

$$\int_0^{2\pi k} \sin \omega t \, dt = 0 \tag{4.54}$$

Now, if you multiply by a constant, on both sides of the integral, the result is still the same:

$$\int_0^{2\pi k} A \sin \omega t \, dt = A \int_0^{2\pi k} A \sin \omega t \, dt = 0 \tag{4.55}$$

That is, the amplitude of the waveform may be larger or smaller, but the average value is still zero.

Now we come to the interesting part. What if we put in, not a constant, but some function of time? That is:

$$\int_0^{2\pi k} g(t)\sin \omega t\ dt = ? \tag{4.56}$$

The answer naturally depends upon what our function of $g(t)$ is. But as we saw in the last chapter, we really only need to worry about sinusoidal functions for $g(t)$. We can extend our analysis to other waveforms by simply considering the Fourier representation of the waveform. Let's look at the specific case where $g(t) = \sin \eta t$.

$$\int_0^{2\pi k} \sin \eta t \sin \omega t = 0, \eta \neq \omega \tag{4.57}$$

Equation 4.57 is called the *orthogonality of sines*. It tells us that, as long as the two sinusoids do not have the same frequency, then the integral of their products will be equal to zero. This may be a little hard to visualize. If so, think back to Equation 4.55. When the frequencies are not the same, the amplitude of the resulting waveform will tend to be *symmetrically* pushed both above and below the x-axis. This may result in some strange-looking waveforms but, over time, the average will come out to zero. In effect, even though $g(t)$ is a function of time, it will have the same effect as if it were the simple constant A.

So what about the case when $\eta = \omega$? If we substitute ω for η in Equation 4.57:

$$\int_0^{2\pi k} \sin \omega t \sin \omega t dt = \int_0^{2\pi k} \sin^2 \omega t dt \neq 0 \tag{4.58}$$

That is, we get the sum of the *square* of the sine wave. When we square the sine waveform, we get a figure like the one shown in Figure 4.15. Since a negative value times a negative value gives a positive value, the negative portion of the original sine wave is moved vertically above the x-axis. The resulting waveform is always positive, so its average value will not be zero.

So far the discussion has made use of analytical functions which are useful in developing algorithms and theoretical concepts. As a practical matter, however, in DSP work we are generally more interested in testing a sequence of numbers (the sampled signal) for orthogonality. At this point, we need to take a slight diversion through the subject of continuous functions versus discrete sequences.

CONTINUOUS FUNCTIONS VS. DISCRETE SEQUENCES

When we look at a function like $y(t) = \sin(2\pi ft)$ we normally think of it as a continuous function of t. If we were to graph the function, we would compute a reasonable number of points and then plot these points. Next, we would draw

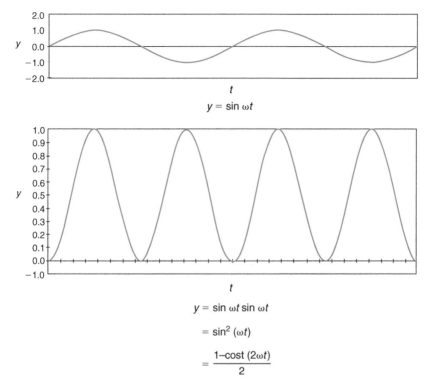

FIGURE 4.15 The average of the square of a sine wave is greater than zero

a continuous and smooth line through all of the points. We would therefore have a continuum of points for *t*, even though we computed the value of the function at a finite number of discrete points.

In general, we can apply numerical techniques to compute a value for any specific function. For example, even if we cannot analytically solve an integral, we can still compute a specific value for it. From Equation 4.16:

$$\int f(x)dx \approx \sum f(x)\Delta x \qquad (4.59)$$

We point this out because it would seem reasonable, when dealing with DSP functions, to adopt the same computational methods. Interestingly enough, we generally do *not*. This fact is not usually emphasized in most texts on DSP, and it can lead to some confusion.

While there is not normally a large leap between continuous and discrete functions in mathematics, it often appears that there is some mysterious difference between discrete and continuous functions in DSP.

FAQs

How and why are discrete and continuous forms of functions different in DSP applications?

In Equation 4.59 we can think of both sides of the equation as finding the area under the curve f. Whether or not we find this area by analytically solving the integral, and then evaluating the resulting function, or by numerically evaluating the right-hand side, we expect to get *essentially* the same answer.

Most DSP applications involve an intensive amount of computation. Anything that can be done to save computation effort is important. Furthermore, it turns out that we are often only interested in *relative* values. In most DSP applications the Δx term is really just a scale factor. For these reasons, we often drop the multiplication by Δx. Thus, it is common to see things like:

$$y_c = \int f(x)dx \quad \text{(the continuous form)}$$

and

$$y_d = \sum f(x) \quad \text{(the discrete form)}$$

Now, these two forms will *not* give us numerically equivalent results. However, surprisingly often, we don't really care. We will demonstrate this concept next as we develop the idea of orthogonality for discrete sequences.

ORTHOGONALITY CONTINUED

The discrete form of Equation 4.56 is generally written as:

$$\sum_{n=-\infty}^{\infty} x[n]\sin\left(2\pi f\,\frac{n}{N}\right) = 0, \text{ if } x[n] \neq \sin\left(\frac{2\pi fn}{N}\right) \tag{4.60}$$

What is the significance of all this? Well, it provides us with a means of testing to see if the sequence $x[n]$ was generated from $\sin(2\pi fn/N)$. This may not seem particularly useful, and in fact, in this form it is *not* particularly useful. This is the case because we need to know the exact phase of $x[n]$ to make Equation 4.60 work. If we could remove this restriction, then Equation 4.60 would have more utility. It would allow us to test to see if the sequence $x[n]$ contained a frequency component at the frequency f. (The importance of this will be made clear in the next chapter, on transforms.)

We would now like to remove the requirement that $x[n]$ be in phase with the sine function. This is where our next key building block comes into play: *quadrature*.

QUADRATURE

The term quadrature has a number of meanings. For our purposes the term is used to refer to signals that are 90° out of phase with each other. The classic example of a quadrature signal is the complex exponential:

$$e^{j\omega} = \cos \omega + j \sin \omega$$

This suggests that the complex exponential may be useful in our quest to come up with a more usable form of Equation 4.60. If we multiplied the sequence $x[n]$ by the complex exponential instead of just the sine function, then we would have a complex sequence. Since a complex number has both phase and magnitude, this allows us much more flexibility in dealing with the phase of the sequence $x[n]$.

To illustrate this concept, take a look at Figure 4.16. The first of three possible phase relationships for the sequence $x[n]$ is shown. In this case the sequence $x[n]$ is in phase with the imaginary part of $e^{j\omega}$. Figure 4.16a shows the imaginary part, and Figure 4.16b shows the real part of $e^{j\omega}$. Figure 4.16c is the function for the sequence:

$$x[n] = \sin\left(\frac{\omega n}{N}\right) \tag{4.61}$$

Now comes the interesting part. Multiplying Figure 4.16a by Figure 4.16c point by point and summing yields:

$$\sum x[n]\text{Im}\left(e^{j\omega n/N}\right) > 0 \tag{4.62}$$

and the real part is:

$$\sum x[n]\text{Re}\left(e^{j\omega n/N}\right) = 0 \tag{4.63}$$

We can see this by simply looking at the graphs in Figure 4.16d and Figure 4.16e. In Figure 4.16d we see two interesting features. First, the frequency has doubled. This is not particularly relevant to our current argument, but it is a nice check: from any trigonometry book we know that squaring a sine wave should double the frequency. The second, and more relevant, point is that the waveform is offset above the x-axis. This means that the waveform has some average value greater than zero.

In Figure 4.16e we see that the waveform is symmetrical about the x-axis. Thus, the average value is zero for the real product.

Figure 4.17 shows the opposite case. In this case, our input function (Figure 4.17c) is:

$$x[n] = \cos\left(\frac{\omega n}{N}\right) \tag{4.64}$$

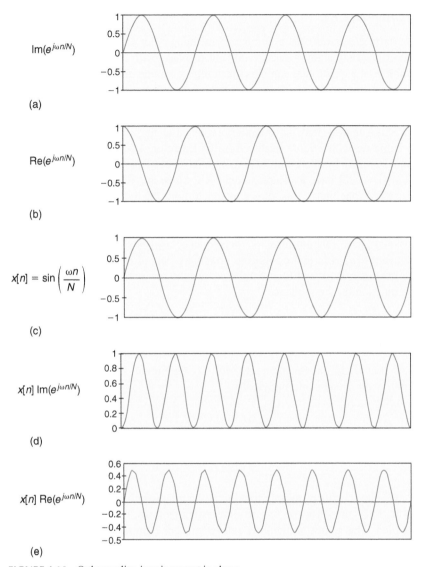

FIGURE 4.16   Orthogonality: imaginary part in phase

The sequence $x[n]$ is in phase with the real part of $e^{j\omega}$. In this case:

$$\sum x[n]\mathrm{Re}\left(e^{j\omega n/N}\right) > 0 \qquad (4.65)$$

as shown in Figure 4.17e.

Now, the *really* interesting part of all of this is shown in Figure 4.18. In this case, the sequence $x[n]$ is 45° (or, equivalently, $\pi/4$ radians) out of phase with

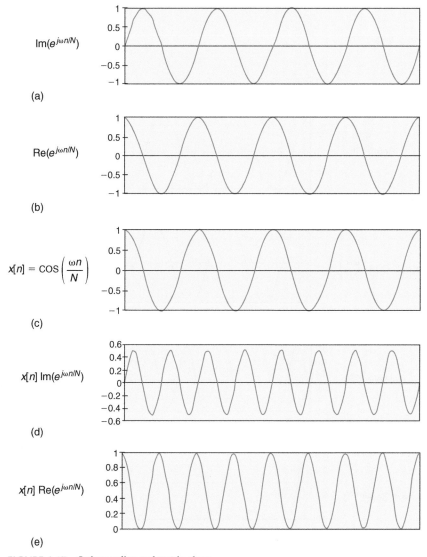

$\text{Im}(e^{j\omega n/N})$

(a)

$\text{Re}(e^{j\omega n/N})$

(b)

$x[n] = \cos\left(\dfrac{\omega n}{N}\right)$

(c)

$x[n]\,\text{Im}(e^{j\omega n/N})$

(d)

$x[n]\,\text{Re}(e^{j\omega n/N})$

(e)

FIGURE 4.17   Orthogonality: real part in phase

both the real and imaginary parts of $e^{j\omega}$. At first, this may seem a lost cause. However, in this case, the $x[n]$ lies in the first quadrant. Therefore, a portion of the signal will be mapped into the real sum of the products and a portion of the signal will be mapped into the imaginary portions of the sum of the products, as shown in Figure 4.18d and Figure 4.18e.

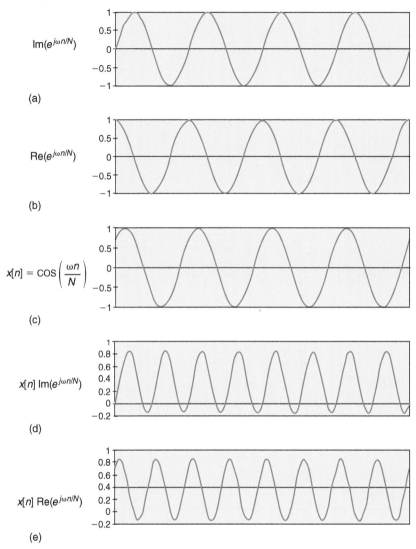

FIGURE 4.18   Orthogonality: quadrature

Figure 4.18e clearly shows this. Each has a value less than the equivalent case when the input signal was in phase with the real or imaginary part. On the other hand, the value is clearly greater than zero.

We are really only interested in the magnitude of the signal, however, so we can take the absolute value of the sum:

$$\left| \sum x[n] e^{j\omega n/N} \right| > 0 \tag{4.66}$$

> **Key Concept**
>
> The key point here is that the magnitude of the complex sum is the same regardless of the phase of $x[n]$ with respect to $e^{j\omega}$.

To summarize what we have just done, if we multiply a sinusoidal signal by another sinusoidal signal of the same *frequency* and *phase,* we can tell if two frequencies are the same. We can tell this because the average value of the product will be greater than zero. (OK, we could tell that just by looking at the two signals, too.)

We can eliminate the problem with the phase by multiplying the input function by the complex exponential. When we do this, it does not matter what the phase of the input signal is: part of the signal will map into the real product, and part of the signal will map into the imaginary product. By taking the absolute value of the complex product, we get the same value as if the signal were in phase with one of the real or imaginary parts.

INSTANT SUMMARY

This chapter has discussed a number of mathematical relationships that are used extensively in digital signal processing.

In DSP, **complex numbers** are of practical importance: they are at the heart of many key DSP algorithms. There is, however, nothing magical about complex numbers. If we remember a couple of simple relationships, complex numbers can be handled as easily as any other number.

We also introduced the concepts of **analog and digital frequencies**. The two are, of course, closely related. At the same time, they are strangely independent of each other. The analog frequency is often dropped in DSP calculations and the digital frequency used instead. Then, in the final result, the analog frequency is restored by scaling the digital frequency. (Often this operation is left out in the discussion—a fact that can be very confusing.)

Next, we demonstrated how a **low-pass filter** and a **high-pass filter** can be developed from a heuristic standpoint. Then, we presented one of the basic concepts needed to develop more sophisticated filters: **convolution**.

The **Fourier series** tells us that any practical signal can be represented as a series of sine waves. This allows us to do all of our analysis of systems using only sinusoidal inputs—a very significant simplification! By looking at the harmonics of any signals that we wish to understand, we can gain a good understanding of the bandwidth requirements for our system. This analysis allows us to specify the sampling rate and the practical frequency cutoffs necessary to implement a practical system.

Orthogonality, as it applies to most DSP work, simply means that multiplying two orthogonal sequences together and taking the sum of the resulting sequence yields a result that is zero. If the multiplication and addition is done numerically, the result may not be *exactly* zero, but it will be close to zero with respect to the amplitude of the functions.

Orthogonality suggests some useful applications. By itself, however, the orthogonality of real functions is of limited value because of an implicit assumption that the two functions (or sequences) are in phase with respect to each other. By using sequences of complex numbers, however, we can bypass the requirement that the functions be in phase. The use of complex numbers in this way is often referred to as **quadrature**.

Transforms

Definitions

In this section we will look at what transforms are and why they are of interest. We will then use the previous discussion on orthogonality and quadrature to develop some useful transforms and their applications. First, we will define some terms used in this chapter. A *transform* is a procedure, equation, or algorithm that changes one group of data into another group of data. The *discrete Fourier transform* (DFT) is a computational technique for computing the transform of a signal. It is normally used to compute the spectrum of a signal from the time domain version of the digitized signal. The *Fourier transform* is a mathematical transform using sinusoids as the basis function. The *z-transform* is a mathematical method used to analyze discrete systems; it changes a signal in the time domain into a signal in the z-domain. The *fast Fourier transform* (FFT) is a very efficient algorithm for calculating the discrete Fourier transform.

BACKGROUND

In general, a mathematical *transform* is exactly what the name implies: it transforms an equation, expression, or value into another equation, expression, or value. One of the simplest transforms is the logarithmic operation. Let's say, for example, that we want to multiply 100 by 1,000. Obviously the answer is 100,000. But how do we arrive at this? There are two approaches. First, we could have multiplied the 100 by 1000. Or we could have used the logarithmic approach:

$$100 \times 1000 = 10^2 \times 10^3 = 10^5$$

The advantage of using the logarithmic approach is, of course, that we only need to add the logarithms (2 + 3) to get the answer. No multiplication is required.

What we have done is use logarithmic operations to *transform* the numbers 100 and 1000 into exponential expressions. In this form we know that addition of the exponents is the same as multiplying the original numbers. This is typically why we perform transforms: the transformed values are, in one way or another, easier to work with.

Another common transform is the simple frequency-to-period relationship:

$$f = 1/P$$

This states that if we know the fundamental period of a signal, we can compute its fundamental frequency—a fact often used in electronics to convert between frequency and wavelength:

$$L = P\lambda$$

where L is the wavelength and λ is the speed of light.

The frequency of a radio wave and its wavelength represent the same thing, of course. But for some things, such as antenna design, it is much easier to work with the wavelength. For others, such as oscillator design, it is simpler to work with the frequency. We commonly transform from the frequency to the wavelength, and the wavelength to the frequency, as the situation dictates.

This leads us to one of the most common activities in DSP: transforming signals. Let's start by looking at a simple example.

Figure 5.1a shows a simple oscillator. If we look at the output of the oscillator as a function of *time*, we would get the waveform shown in Figure 5.1b.

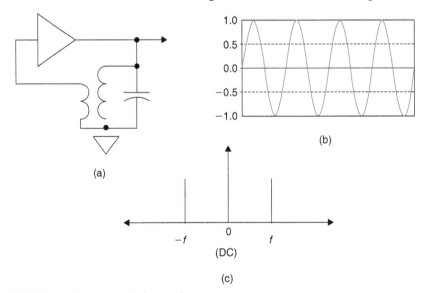

FIGURE 5.1   Spectrum analysis example

If we look at the output as a function of *frequency*, we would get the result shown in Figure 5.1c. Notice that in Figure 5.1c we have shown both the positive frequency *f* and the negative frequency −*f*.

> **Insider Info**
>
> *In most electronics applications, we don't normally show the negative frequency spectrum. The reason for this is that, for any real-valued signal, the spectrum will be symmetrical about the origin. Notice that in Figure 5.1c we can determine both the frequency and the amplitude of the signal. We get the frequency from the distance from the origin and, of course, we get the amplitude from the position on the y-axis.*

In this simple case, it was easy to move from the *time domain* (Figure 5.1b) of a signal to the *frequency domain* (Figure 5.1c) because we know the simple relationship:

$$f = 1/P$$

Now, what if we wanted to look at the spectrum of a more complicated signal—for example, a square wave?

We can do this by inspection from our work on the Fourier series. We know that a square wave is composed of a sine wave at the fundamental frequency, and a series of sine waves at harmonic frequencies. With this information, we can take a signal like the one in Figure 5.2a and find its spectrum. The spectrum is shown in Figure 5.2b.

This process of converting from the time domain to the frequency domain is called a *transform*. In this case, we have performed the transform heuristically,

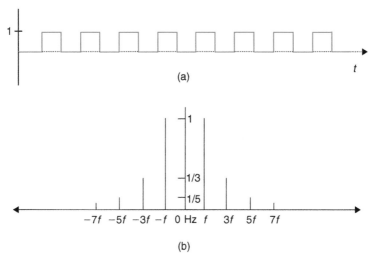

FIGURE 5.2 Transform of a square wave

using the knowledge we have already developed of the square wave. There are lots of applications for transforms. Often, it is impossible to tell what frequency components are present by simply looking at the time domain representation of a signal. If we can see the signal's spectrum, however, these frequency components become obvious. This has direct application in seismology, radar and sonar, speech analysis, vibration testing, and many other fields.

With all of these applications, it is only logical to come up with some general-purpose method for transforming a signal from the time domain to the frequency domain (or vice versa). Fortunately, it turns out that there is a relatively simple procedure for doing this. As you have probably already guessed, it makes use of the techniques from the last chapter: quadrature and orthogonality. Before we move on, however, we need to take a detour through another interesting tool: the z-transform.

THE Z-TRANSFORM AND DFT

In Chapter 4 we reviewed the Taylor series for describing a function. In that discussion, we pointed out that virtually any function can be expressed as a polynomial series. The z-transform is a logical extension of this concept.

We will start by looking at the variable z, and the associated concept of the z-plane. Next, we will give the definition of the z-transform. We will then take a look at the z-transform in a more intuitive way. Finally, we will use it to derive another important (and simpler) transform: the *discrete Fourier transform* (DFT).

Insider Info

The Fourier transform family consists of four categories of transforms; which one is used depends on the type of signal encountered. The categories are called Fourier transform, Fourier series, discrete Fourier transform, and discrete time Fourier transform. These names have evolved over a long time and can be very confusing. The discrete Fourier transform is the one that operates on a periodic sampled time domain signal, and is the one that is most relevant to DSP.

The variable z is a complex quantity. As we saw in Chapter 4, there are a number of ways of expressing a complex number. While all of the methods are interchangeable, some work better in certain situations than others, and the z-transform is no exception. Thus, the variable z is normally defined as:

$$z = re^{j\omega} \tag{5.1}$$

In words, any point on the z-plane can be defined by the angle formed by $e^{j\omega}$, located r units from the origin. Or, more succinctly, the point P is a function of the variables r and ω. This concept is shown graphically in Figure 5.3.

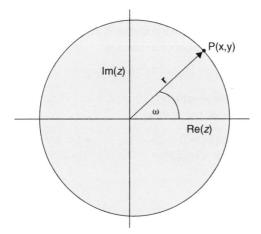

FIGURE 5.3 The z-plane

Now, let's look back at the Taylor series:

$$f(x) = \sum_{n=0}^{\infty} a_n x^n$$

This is a real-valued function that expresses the value of $f(x)$ in terms of the coefficients a_n, and the variable x raised to a corresponding power. With only minimal effort, we can generalize this expression to a complex form using Equation 5.1:

$$f(z) = \sum_{-\infty}^{\infty} a_n z^n \qquad (5.2)$$

where a_n is the input sequence.

Interesting, but what does this have to do with signal processing? Well, as we have seen so far we are normally dealing with signals as sequences of discrete values. It turns out that there are some analytical advantages to using negative values for n, but otherwise it does not make any difference to the overall discussion.

For example, let's say we have an input sequence:

$$a[n] = \{3, 2, 1\}$$

We could express this sequence, using Equation 5.2, as:

$$f[z] = 3z^0 + 2z^{-1} + 1z^{-2} \qquad (5.3)$$

Now, *why* we would want to do this probably isn't clear, but we will get to this in a minute. In the meantime, let's look at one of the often cited attributes of the z-transform. There is a very interesting property of a series called the shifting property. For example, we could shift the sequence $x[n]$ to the sequence $x[n + 1]$. This would then produce a function:

$$g[z] = 3z^1 + 2z^0 + z^{-1} \tag{5.4}$$

Obviously $f[z]$ is not equal to $g[z]$.
For example, if we let $z = 2$, then:

$$f[2] = 3 \times 2^0 + 2 \times 2^{-1} + 1 \times 2^{-2}$$
$$= 4.25$$

and:

$$g[2] = 3 \times 2^1 + 2 \times 2^0 + 1 \times 2^{-1}$$
$$= 8.5 \tag{5.5}$$

If we look at these two values we might notice that $y[2]$ is equal to half the value of $y[2]$. And, not coincidentally, z^{-1} is also equal to 0.5. In fact, in general:

$$y[z] = z^{-1}G[z + 1]$$

where the capital letter indicates the z-transform expression of the function. The relationship demonstrated in Equation 5.5 is called the *shifting theorem.*

The shifting theorem is not as mysterious as it might seem at first glance if we remember that multiplying by variables with exponents is accomplished by adding the exponents. Thus, multiplying by z^{-1} is really the same as decrementing the exponent by 1. Indeed, the exponent is often viewed as the index of the sequence—just like a subscript.

Key Concept

The shifting theorem plays an important role in the analytical development of functions using the z-transform. It is also common to see the notation z^{-1} used to indicate a delay. We will revisit the shifting theorem when we look at the expression for the IIR filter.

Now, for a more direct application of the z-transform. As we mentioned earlier, we can think of z as a function of the frequency ω and magnitude r. If we set $r = 1$, then Equation 5.2 reduces to:

$$Y(z) = \sum_{n=-\infty}^{\infty} a_n z^{-n}, \quad \text{letting} \quad r = 1$$

$$Y[e^{-j\omega}] = \sum_{n=-\infty}^{\infty} a_n e^{-j\omega n/N} \tag{5.6}$$

The left side of Equation 5.6 is clearly an *exponential* function of the frequency ω. This has two important implications. First, a graph of Y as a function is nearly impossible: it would mean graphing a complex result for a complex variable, requiring a four-dimensional graph. A second consideration is that, effectively, the expression $Y[e^{-j\omega}]$ maps to the unit circle on the z-plane. For example, if we have $\omega = 0$:

$$Y[e^{-j\omega}] = Y[\cos 0 + j\sin 0] = Y[1, 0]$$

or if $\omega = \pi/4$, then

$$Y[e^{-j\omega}] = Y\left[\cos\frac{\pi}{4} - j\sin\frac{\pi}{4}\right] = Y\left[\frac{\sqrt{2}}{2}, \frac{\sqrt{2}}{2}\right]$$

In our discussion of orthogonality, we pointed out that the function Y, because it is complex, has information about both the phase and magnitude of the spectrum in the signal. Sometimes we care about the phase, but often we do not. If we do not care about the phase, then we get the amplitude by taking the absolute value of Y.

We can make a further simplification to Equation 5.6. It is acceptable to drop the $e^{-j\omega}$ term and express Y simply as a function of ω. Therefore, we generally express Equation 5.6 as:

$$Y(\omega) = \sum_{n=-\infty}^{\infty} x[n]e^{-j\omega n/N} \tag{5.7}$$

Believe it or not, we are actually getting somewhere. Notice that the right side of Equation 5.7 is familiar from our discussion of orthogonality. With this revelation we can translate the action of Equation 5.7 into words:

Let's assume we have an input signal sequence $\{x[n]\}$. We can determine if the signal has a frequency component at the frequency ω by evaluating the sum in Equation 5.7. If we do this for values of ω ranging from $-\pi$ to π we will get the complete spectrum of the signal.

Key Concept

Equation 5.7, when evaluated at the discrete points $\omega_k = 2\pi k/N$, $k = 0, 1 \dots$ $N - 1$, is commonly called the *discrete Fourier transform* (DFT). It is one of the most common computations performed in signal processing. As we noted above, it allows us to transform a function of time into a function of frequency. Or, equivalently, it means we can see the spectrum of an input signal by running it through the DFT. The DFT can be calculated in several different ways, which we'll discuss as we move through this chapter and the following chapters.

APPLICATION OF THE DFT

We will pull this all together with an example. First, we will generate a signal. Since we are generating the signal we will know its spectrum (it's always nice to know the correct answer before setting out to solve a problem). Next, we will use the DFT to compute the spectrum, and then see if it gives the answer we expect.

For this example, we will set everything up using a spreadsheet. Table 5.1 shows how we generate the signal. It is composed by adding together two separate signals:

$$f_n = \sin\left[\frac{2\pi hn}{N}\right], \ h = 2$$

and

$$g_n = (0.5)\sin\left[\frac{2\pi hn}{N} + \frac{\pi}{4}\right], \ h = 4$$

where h is used to denote the frequency in cycles per unit time. Notice that the first component (f) and the second component (g) are out of phase with each other by $90°$ $(\pi/4)$. This will help illustrate why we need to use complex numbers in the computation.

The resulting waveform is shown in Figure 5.4. In Figure 5.5 we can see the spectrum for the signal. We can, of course, draw the spectrum by simple inspection of the two components. But let's see if the DFT can give us the same information via computation.

In Table 5.2 we have set up the DFT with a frequency of zero. In other words, we are going to see if there is any DC component. As you can see, the real part of the sum is small and the imaginary part of the sum is zero, so of course the absolute value is small. We can repeat this for any frequency other than $f = 2$ or $f = 4$ and we will get a similar result. So let's look at these last two cases.

Tables 5.2, 5.3 and 5.4 are set up to show the index n in the first column. The second column is the signal $f + g$. The third column is $\text{Re}(e^{-j\omega n/N})$, and the fourth column is $\text{Im}(e^{-j\omega n/N})$. The fifth column is $\text{Re}(f_n e^{-j\omega n/N})$. The sixth column is, naturally, $\text{Im}(f_n e^{-j\omega n/N})$.

For $Y[2]$ we would expect to get a large value, since one component of the signal was generated at this frequency. Since the signal was generated with the sine function, we would expect the value to be imaginary. This is exactly what we see in Table 5.3. The value we get is *not* 1, but by convention, when we plot the spectrum we normalize the largest value to 1.

The actual value in Table 5.3 is 16.0. This is a dimensionless number, not really corresponding to any physical value. If we had used a larger number of samples, the number would have been larger. Correspondingly, a smaller number of samples would have given us a smaller value. By normalizing the value,

TABLE 5.1 Signal generation

n	$f = \sin(2\pi(2)n/N)$	$g = \sin(2\pi(4)n/N + \pi/4)/2$	$f + g$
0	0.000	0.354	0.354
1	0.383	0.500	0.883
2	0.707	0.354	1.061
3	6.924	0.000	0.924
4	1.000	−0.354	0.646
5	0.924	−0.500	0.424
6	0.707	−0.354	0.354
7	0.383	0.000	0.383
8	0.000	0.354	0.354
9	−0.383	0.500	0.117
10	−0.707	0.354	−0.354
11	−0.924	0.000	−0.924
12	−1.000	−0.354	−1.354
13	−0.924	−0.000	−1.424
14	−0.707	−0.354	−1.061
15	−0.383	0.000	−0.383
16	0.000	0.354	0.354
17	0.383	0.500	0.883
18	0.707	0.354	1.061
19	6.924	0.000	0.924
20	1.000	−0.354	0.646
21	6.924	−0.500	0.424
22	0.707	−0.354	0.354
23	0.383	0.000	0.383
24	0.660	0.354	0.354
25	−0.383	0.500	0.117
26	−0.707	0.354	−0.354
27	−0.924	0.000	−0.924
28	−1.000	−0.354	−1.354
29	−0.924	−0.500	−1.424
30	−0.707	−0.354	−1.061
31	−0.383	0.000	−0.383
32	6.666	0.354	0.354

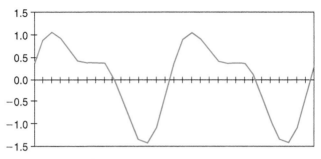

FIGURE 5.4   Composite waveform

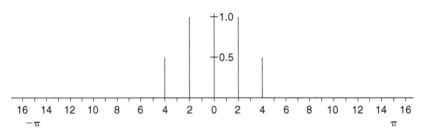

FIGURE 5.5   Spectrum for the signal in Figure 5.4

TABLE 5.2 **DFT with frequency = 0**

n	f + g	cos(2π(0)n/N)	sin(2π(0)n/N)	Real Part	Imag. Part
0	0.354	1.000	0.000	0.354	0.000
1	0.883	1.000	0.000	0.883	0.000
2	1.061	1.000	0.000	1.061	0.000
3	0.924	1.000	0.000	0.924	0.000
4	0.646	1.000	0.000	0.646	0.000
5	0.424	1.000	0.000	0.424	0.000
6	0.354	1.000	0.000	0.354	0.000
7	0.383	1.000	0.000	0.383	0.000
8	0.354	1.000	0.000	0.354	0.000
9	0.117	1.000	0.000	0.117	0.000
10	−0.354	1.000	0.000	−0.354	0.000
11	−0.924	1.000	0.000	−0.924	0.000
12	−1.354	1.000	0.000	−1.354	0.000
13	−1.424	1.000	0.000	−1.424	0.000

TABLE 5.2 (continued)

n	f + g	cos(2π(0)n/N)	sin(2π(0)n/N)	Real Part	Imag. Part
14	−1.061	1.000	0.000	−1.061	0.000
15	−0.383	1.000	0.000	−0.383	0.000
16	0.354	1.000	0.000	0.354	0.000
17	0.883	1.000	0.000	0.883	0.000
18	1.061	1.000	0.000	1.061	0.000
19	0.924	1.000	0.000	0.924	0.000
20	0.646	1.000	0.000	0.646	0.000
21	0.424	1.000	0.000	0.424	0.000
22	0.354	1.000	0.000	0.354	0.000
23	0.383	1.000	0.000	0.383	0.000
24	0.354	1.000	0.000	0.354	0.000
25	0.117	1.000	0.000	0.117	0.000
26	−0.354	1.000	0.000	−0.354	0.000
27	−0.924	1.000	0.000	−0.924	0.000
28	−1.354	1.000	0.000	−1.354	0.000
29	−1.424	1.000	0.000	−1.424	0.000
30	−1.061	1.000	0.000	−1.061	0.000
31	−0.383	1.000	0.000	−0.383	0.000
			sum =	0	0
			abs(sum) =	0	

TABLE 5.3

n	f + g	cos(2π(2)n/N)	sin(2π(2)n/N)	Real Part	Imag. Part
0	0.354	1.000	0.000	0.354	0.000
1	0.883	0.924	0.383	0.815	0.338
2	1.061	0.707	0.707	0.750	0.750
3	0.924	0.383	0.924	0.354	0.854
4	0.646	0.000	1.000	0.000	0.646
5	0.424	−0.383	0.924	−0.162	0.392

TABLE 5.3 (continued)

n	f + g	cos(2π(2)n/N)	sin(2π(2)n/N)	Real Part	Imag. Part
6	0.354	−0.707	0.707	−0.250	0.250
7	0.383	−0.924	0.383	−0.354	0.146
8	0.354	−1.000	0.000	−0.354	0.000
9	0.117	−0.924	−0.383	−0.108	−0.045
10	−0.354	−0.707	−0.707	0.250	0.250
11	−0.924	−0.383	−0.924	0.354	0.854
12	−1.354	0.000	−1.000	0.000	1.354
13	−1.424	−0.383	−0.924	−0.545	1.315
14	−1.061	0.707	−0.707	−0.750	0.750
15	−0.383	0.924	−0.383	−0.354	0.146
16	0.354	1.000	0.000	0.354	0.000
17	0.883	0.924	0.383	0.815	0.338
18	1.061	0.707	0.707	0.750	0.750
19	0.924	0.383	0.924	0.354	0.854
20	0.646	0.000	0.707	0.000	0.646
21	0.424	−0.383	0.924	−0.162	0.392
22	0.354	−0.707	0.707	−0.250	0.250
23	0.383	−0.924	0.383	−0.354	0.146
24	6.354	−1.000	0.000	−0.354	0.000
25	0.117	−0.924	−0383	−0.108	−0.045
26	−0.354	−0.707	−0.707	0.250	0.250
27	−0.924	−0.383	−0.924	0.354	0.854
28	−1.354	0.000	−1.000	0.000	1.354
29	−1.424	0.383	−0.924	−0.545	1.315
30	−1.061	0.707	−0.707	−0.750	0.750
31	−0.383	0.924	−0.383	−0.354	0.146
			sum =	0.000	16
			abs(sum) =	16	

we account for this variation in the signal length. With this caveat in mind, we can think of the normalized value as the amplitude of the signal.

What can we expect for the transform of the second frequency component? Since the first component had a non-normalized value of 16, we would expect the second frequency component to have a value of 8. Further, since the second component was generated with a $\pi/4$ phase shift, we would expect this value to be distributed between the imaginary and the real components.

In Table 5.4 we evaluate $Y[4]$, and we see that we get exactly what we would expect.

In later chapters, we will see additional uses for the DFT. But for now, we'll just look at some characteristics of the DFT.

First, the DFT works in both directions: if we feed the spectrum of a signal into the DFT, we will get the time domain representation of the signal out. We may have to add a scaling factor (since we normalized the DFT). Sometimes the DFT with this normalizing factor is called the *inverse discrete Fourier transform (IDFT)*, (Remember that this inversion applies only to the DFT. It is *not* true for the more general z-transform.)

Next, we'll look at two other transforms: the Fourier transform and the Laplace transform. Both are covered here briefly. We are discussing them primarily to make some important comparisons to the DFT and their general relationship to signal processing.

THE FOURIER TRANSFORM

Considering that we just discussed the discrete Fourier transform, we might gather that the Fourier transform is simply the continuous case of the DFT. One of the confusing things in the literature of DSP is that, in fact, the DFT is *not* simply the numerical approximation of the Fourier transform obtained by using discrete mathematics. This goes back to our previous discussion about continuous versus discrete functions in DSP.

Insider Info

This is why we approached the DFT via the z-transform. It really is a special case of the z-transform, and therefore the derivation is more direct. In the DFT, as in the z-transform (or any power series representation), we are working with discrete values of the function. When we move to the continuous case of the Fourier transform, we are actually working with the integral of the function. Geometrically, this can be thought of as follows: The discrete form uses points on the curve of a function. The continuous form makes use of the area under the curve. In practice, the distinction is not necessarily critical. But it can lead to some confusion when trying to implement algorithms from the literature, or when studying the derivation of certain algorithms.

TABLE 5.4

n	f + g	cos(2π(4)n/N)	sin(2π(4)n/N)	Real Part	Imag. Part
0	0.354	1.000	0.000	0.354	0.000
1	0.883	0.707	0.707	0.624	0.624
2	1.061	0.000	1.000	0.000	1.061
3	0.924	−0.707	0.707	−0.653	0.653
4	0.646	−1.000	0.000	−0.646	0.000
5	0.424	−0.707	−0.707	−0.300	−0.300
6	0.354	0.000	−1.000	0.000	−0.354
7	0.383	0.707	−0.707	0.271	−0.271
8	0.354	1.000	0.000	0.354	0.000
9	0.117	0.707	0.707	0.083	0.083
10	−0.354	0.000	1.000	0.000	−0.354
11	−0.924	−0.707	0.707	0.653	−0.653
12	−1.354	−1.000	0.000	1.354	0.000
13	−1.424	−0.707	−0.707	1.007	1.007
14	−1.061	0.000	−1.000	0.000	1.061
15	−0.383	0.707	−0.707	−0.271	0.271
16	0.354	1.000	0.000	0.354	0.000
17	0.883	0.707	0.707	0.624	0.624
18	1.061	0.000	1.000	0.000	1.061
19	0.924	−0.707	0.707	−0.653	0.653
20	0.646	−1.000	0.000	−0.646	0.000
21	0.424	−0.707	−0.707	−0.300	−0.300
22	0.354	0.000	−1.000	0.000	−0.354
23	0.383	0.707	−0.707	0.271	−0.271
24	0.354	1.000	0.000	0.354	0.000
25	0.117	0.707	0.707	0.083	0.083
26	−0.354	0.000	1.000	0.000	−0.354
27	−0.924	−0.707	0.707	0.653	−0.653
28	−1.354	−1.000	0.000	1.354	0.000
29	−1.424	−0.707	−0.707	1.007	1.007
30	−1.061	0.000	−1.000	0.000	1.061
31	−0.383	0.707	−0.707	−0.271	0.271
			sum =	5.657	5.657
			abs(sum) =	8	

The forms of the DFT and the Fourier transform are quite similar. The Fourier transform is defined as:

$$H(\omega) = \int_{-\infty}^{\infty} f(t)e^{-j\omega t} dt \tag{5.8}$$

The Fourier transform operator is often written as F:

$$H(\omega) = F(f(t))$$

or, equivalently:

$$x(t) \Leftrightarrow X(\omega)$$

It is a fairly uniform convention in the literature to use lower-case letters for time domain functions and uppercase letters for frequency domain functions. In this book, this convention is followed.

PROPERTIES OF THE FOURIER TRANSFORM

Table 5.5 presents a table of the common mathematical properties of the Fourier transform. These properties follow in straightforward fashion from Equation 5.8. For example, Property 1 states that:

$$aH(\omega) = a \int_{-\infty}^{\infty} f(t)e^{-j\omega t} dt = F(af(t))$$

where a is an arbitrary constant.

It is worth noting that, as with the geometric series discussed in Chapter 4, the shifting operation applies to the Fourier transform:

$$x(t - \tau) \Leftrightarrow e^{-j\omega\tau} X(\omega)$$

This property is rarely used with relationship to the Fourier transform. It is pointed out here because of the significance it plays in the relationship to the z-transform discussion presented earlier.

A number of other properties of the Fourier transform are pointed out in Table 5.5. Some of these properties, such as the homogeneity property discussed above, follow fairly naturally. Other properties, such as convolution, have not yet been discussed in a context that makes sense. These properties will be discussed in later chapters.

THE LAPLACE TRANSFORM

The Laplace transform is a natural extension of the Fourier transform. Typically, the Laplace transform does not play a direct role in DSP applications. However, it is being discussed here for several reasons.

TABLE 5.5 Some properties of the Fourier transform

Property	Time function f(t)	Fourier transform X(ω)
1 Homogeneity	$ax(t)$	$aX(\omega)$
2 Additivity	$x(t) + y(t)$	$X(\omega) + Y(\omega)$
3 Linearity	$ax(t) + by(t)$	$aX(\omega) + bY(\omega)$
4 Differentiation	$\dfrac{d^n}{dt^n} x(t)$	$(j\omega)^n X(\omega)$
5 Integration	$\displaystyle\int_{-\infty}^{t} x(t)dt$	$\dfrac{X(\omega)}{j\omega} + \dfrac{1}{2} X(0)\delta(f)$
6 Sine Modulation	$x(t)\sin(\omega_0 t)$	$\dfrac{1}{2}[X(\omega - \omega_0) + X(\omega + \omega_0)]$
7 Cosine Modulation	$x(t)\cos(\omega_0 t)$	$\dfrac{1}{2}[X(\omega - \omega_0) - X(\omega + \omega_0)]$
8 Time Shifting	$X(t - \tau)$	$e^{-j\omega\tau}X(\omega)$
9 Time Convolution	$\displaystyle\int_{-\infty}^{\infty} h(t - \tau)x(\tau)dt$	$H(\omega)X(\omega)$
10 Multiplication	$x(t)h(t)$	$\displaystyle\int_{-\infty}^{\infty} X(\omega)Y(\omega - \lambda)d\lambda$
11 Time and Frequency Scaling	$x\left(\dfrac{1}{a}\right), a > 0$	$a(Xa\omega)$
12 Duality	$X(t)$	$x(-f)$
13 Conjugation	$x^*(t)$	$X^*(-f)$

One reason is simply to provide completeness of the discussion of transforms in general. Another is the fact that the Laplace transform is often used in many electronics applications that have analogous DSP operations. For example, analog filters are often evaluated using the Laplace transform.

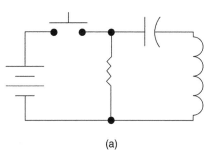
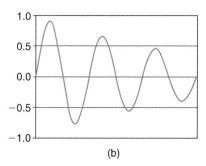

(a)   (b)

FIGURE 5.6   Damped LRC circuit

FAQs

Why do we need to go beyond the Fourier transform?

As noted earlier, the Fourier transform can be used to generate almost any wave-form from a series of sinusoidal signals. Some signals, however, are either diffi-cult or mathematically impossible to model efficiently. Consider, for example, the case of an LRC circuit, as shown in Figure 5.6. The general response of this circuit is a second-order differential equation:

$$L\frac{d^2q}{dt^2} + R\frac{dq}{dt} + \frac{q}{C} = v \tag{5.9}$$

$V(t)$ will have the general solution:

$$Ke^{\tau}e^{j\omega} \tag{5.10}$$

The circuit response for the underdamped case is also shown in Figure 5.6. Notice that Equation 5.10 simply states what most electrical engineers know intu-itively: the response is a damped sine wave. Mathematically, that is a sinusoid multiplied by an exponential function of time. In other words, the output will sim-ply be a "ringing waveform"—a sine wave whose amplitude diminishes exponen-tially over time.

Solving (or mathematically modeling) something like this with the Fourier transform quickly becomes difficult. The sinusoidal components of the Fourier series are all uniform in amplitude over time. This, naturally, suggests that we expand our definition of the Fourier transform to include an expression something like the one shown in Equation 5.10. This gives us:

$$L(x(t)) = \int_0^\infty x(t)e^{\alpha t}e^{-j\omega t}dt \tag{5.11}$$

Notice that Equation 5.11 is just our definition of the Fourier transform with the addition of the $e^{\alpha t}$ term. In fact, if you set α equal to zero, then Equation 5.11 reduces back to the Fourier transform. Generally, Equation 5.11 is simplified by defining a complex variable $s = \alpha + j\omega$. With this substitution, Equation 5.11 then becomes:

$$L(x(t)) = X(s) = \int_0^\infty x(t)e^{-st}dt \qquad (5.12)$$

This is the classic definition of the Laplace transform. One very interesting aspect of the Laplace transform is that it provides a handy means of solving differential equations, analogous to using logarithms to perform multiplication by adding the exponents.

- First, the individual functions are converted to an expression in the variable s via the Laplace transform.
- Next, the overall system equation is solved *algebraically,*
- Then, the solution is converted back from a variable in s to a variable in t by the inverse Laplace transform.

For example, an inductor become sL, and a capacitor becomes $1/sC$. The loop equation for the circuit shown in Figure 5.6 then can be expressed as:

$$sLI(s) + RI(s) + \frac{1}{Cs}I(s) = V(s) \qquad (5.13)$$

Equation 5.13 is mathematically equivalent to Equation 5.9. Notice, however, that Equation 5.13 is an algebraic expression; there are no differential operators required.

As we noted earlier, the Laplace transform is not often a direct player in DSP applications. Therefore, the development here is kept very brief. In future chapters, however, we will occasionally return to the Laplace transform to make some comparisons and analogies, and to remove some points of confusion between the Laplace transform and the z-transform.

FAST FOURIER TRANSFORM (FFT)

Unfortunately, the number of complex computations needed to perform the DFT is proportional to N^2. Calculations can take a long time. The *fast Fourier transform* (FFT) refers to a group of clever algorithms, all very similar, that uses fewer computational steps to efficiently compute the DFT. Reducing the number of computational steps is of course important if the transform has to be computed in a real-time system. Fewer steps implies faster processing time, and higher sampling rates are possible.

From a purely mathematical point of view, DFT and FFT do the same job. The FFT becomes more efficient when the FFT point size increases to several thousand. If only a few spectral points need to be calculated, the DFT may actually be more efficient.

Although the FFT requires just a few lines of computer code, it is a complicated algorithm. However, be reassured that DSP designers often use published FFT routines without completely understanding their inner workings.

Insider Info

The FFT was popularized by J.W. Cooley and J.W. Tukey in the 1960s. It was actually a rediscovery of an idea of Runge (1903) and Danielson and Lanczos (1942), first occurring prior to the availability of computers and calculators, when numerical calculation could take many man-hours. In addition, the German mathematician Karl Friedrich Gauss (1777–1855) had used the method more than a century earlier. (from Kester, Mixed-Signal and DSP Design Techniques, *Elsevier, 2003)*

INSTANT SUMMARY

In this chapter the concept of orthogonality and quadrature have been developed into the *discrete Fourier transform* (DFT). From there, we moved to the *Fourier transform*. The Fourier transform was shown to map a function of time into a function of frequency. This is just the mathematical equivalent of a spectrum analyzer. The Fourier transform was then expanded into the *Laplace transform*. A more efficient way to calculate the DFT was then discussed, the *fast Fourier transform* (FFT).

These methods will be described in more detail in the next chapters.

Digital Filters

Definitions

In the previous chapters we developed a number of tools for working with signals. In order to keep the discussion as tight as possible, these tools were generally presented in a context where they could be understood independently. *Convolution*, for example, was presented as a generalization of the moving average filter. In a similar manner, the *DFT* was shown to be a tool that mapped a function of time (the signal) to a function of frequency (the signal's spectrum). We also pointed out, though we did not demonstrate it, that the DFT was a reversible function: given a signal's spectrum, we could use the DFT to get the signal.

It is now time to start tying these tools together to develop a more sophisticated methodology for filter design. First, let's look at some definitions of terms that we'll encounter. A *Finite Impulse Response filter* (FIR) is a filter whose architecture guarantees that its output will eventually return to zero if the filter is excited with an impulse input. FIR filters are unconditionally stable. An *Infinite Impulse Response filter* (IIR) is a filter that, once excited, may have an output for an infinite period of time. Depending on a number of factors, an IIR may be unconditionally stable, conditionally stable, or unstable. A *window*, as applied to DSP, refers to a special function that shapes the transfer function. It is typically used to tweak the coefficients of filters.

FIR FILTERS

Normally, we think of a filter as a function *of frequency*. That is, we draw a graph showing what frequencies we want to let through and what frequencies we want to block. Such graphs are shown in Figure 6.1, where we show the three most common types of filters: the low-pass, bandpass, and high-pass filter.

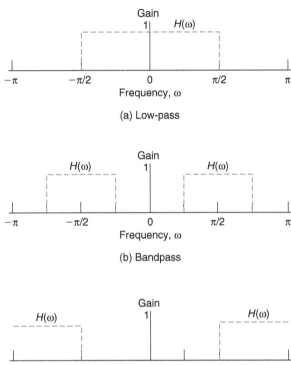

FIGURE 6.1   Three standard filters

In Chapter 4 we looked at the simple moving average filter. We saw how we could implement it as a convolution of the input signal $x[n]$ with the filter function $h[k]$, where $h[k] = 1/k$. We found $h[k]$ by a purely intuitive process. However, we could also find the function $h[k]$ directly from the DFT.

This provides us with a simple and direct way of generating a filter: we define a filter as a function of frequency $H[\omega]$. We then use the DFT to convert $H[\omega]$ to the sequence $h[k]$. Convolving $h[k]$ with $x[n]$ will then give us our filter output $y[n]$! This is another way of looking at the corollary that convolution in the time domain is equivalent to multiplication in the frequency domain. We will look at a practical example, using the DSP *Calculator* software, shortly. First, however, let's point out that a filter of this type is called a *Finite Impulse Response* filter, or FIR. Let's explore a few of the characteristics of the FIR.

What Is an FIR Filter?

The simplest example of a causal FIR filter is our simple moving average filter. As we noted in Chapter 5, the moving average filter can be generated by convolving the input sample x[n] with the transfer function $h[n]$. In the general form, an FIR filter then is:

$$y(n) = \sum_{n=0}^{L} h(m)x(n-m) \qquad (6.1)$$

where L is the length of the filter, and m and n are indexes.

Technology Trade-offs

The FIR filter has several advantages. Since it does not have any poles, it is always guaranteed to be stable. Another advantage is that if the weights are chosen to be symmetrical, the filter has a *linear-phase response*—that is, all frequency components experience the same time delay through the filter. There is no risk of distortion of compound signals due to phase shift problems. Further, knowing the amplitude of the input signal, $x(n)$, it is easy to calculate the maximum amplitude of the signals in different parts of the system. Hence, numerical overflow and truncation problems can easily be eliminated at design time.

The drawback with the FIR filter is that if sharp cut-off filters are needed, so is a high-order FIR structure, which results in long delay lines. FIR filters having hundreds of taps are, however, common today, thanks to low-cost integrated circuit technology and high-speed digital signal processors.

FIR filters get their name from—naturally enough—the way they respond to an *impulse*. For our definition, an impulse is an input of value 1 lasting just long enough to be sampled once and only once. If the response of the filter *must* be finite, then the filter is an FIR. From a practical point of view, a finite response means that, when excited by a unit impulse, the filter's output will return to zero in a reasonable amount of time.

Our simple averaging filters are examples of noncausal FIR filters; given an impulse input, the output will eventually return to zero. As long as the response *must* return to zero for an impulse input, the filter is classified as an FIR. The other major type of filter is the *Infinite Impulse Response* (IIR) filter. As we will see, an IIR filter *may* return to zero for an impulse response, but its architecture does not *require* this to happen.

One helpful way of looking at an FIR filter is shown in Figure 6.2. This type of architectural drawing is generally called a *flow diagram*. As the name implies, a flow diagram sketches the flow of the signal through the system. Notice that the input sequence is shown in what may—intuitively—appear to be the reverse order. In practice, this format is simply showing that f_0 is the first sample of the input sequence. The opposite, but more common, convention is used on the output sequence y.

Several other things in Figure 6.2 deserve comment. The square boxes represent multiplication and the arrows represent delay. Each box is commonly called a *tap*. In this drawing, we have been careful to show two outputs. The output on the bottom of the box is the product of the input sequence and $h(n)$.

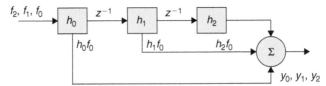

FIGURE 6.2   Standard architecture for an FIR filter

For the first box, and the first computation cycle, this would be $h_0 f_0$. The output from the right side of the box is just the input delayed by one cycle time. The output of the second box would be $h_1 f_0$ after the second cycle of computation.

The symbol z^{-1} is the standard notation for a unit delay. The circle represents summation, and the output of the summation is the output of our filter.

The simple averaging filter from Chapter 4 is implemented by setting $h(n) = 1/3$ for $n = 0, 1, 2$. Notice that the flow diagram then exactly mimics both the simple averaging routine and the more elaborate convolution sum.

> **Insider Info**
>
> *It is worth noting that the flow diagram works equally well for either a software or a hardware implementation. Normally, an FIR filter is implemented in software. However, for systems that require the fastest performance, there is no reason that the multiplication and addition cannot be handled by separate hardware units.*

In the real world, when we sit down to design a filter we are usually most concerned with the frequency response. Other considerations are also important, but they are generally second-order concerns. These additional characteristics include such things as the stability of the filter, phase delay, and the cost of implementing the filter. It is worthwhile to look at these second-order concerns before we proceed to a discussion of designing with FIR filters.

Stability of FIR Filters

One of the great advantages of the FIR filter is that it is inherently stable.

> **Key Concept**
>
> **Regardless of what signal we feed into an FIR filter or how long we feed the signal in, when we set the input to zero the output will eventually go to zero.**

This conclusion becomes obvious when we think through what the filter is doing. Since it is just multiplying and adding up various parts of the input signal, it follows that the products will eventually all be zero after the last element of the input signal propagates through the filter. This also makes it easy to figure out what the worst-case delay through the filter will be. It is simply the number of taps times the sample rate.

As we will see, this inherent stability is not universal to all digital filters.

Cost of Implementation

The cost of implementation is not just a matter of dollars. The cost is also measured in the resources required and in how long it takes these resources to do the job.

For example, as we mentioned earlier, it is possible to improve the response of an FIR filter by simply increasing the number of taps we use. This has several important consequences, however. First, the more taps we use, the longer it takes to compute the output. For a real-time system, this computation must be completed in less than one sample interval. Further, the more taps we use, the greater the phase delay of the filter. Also of concern is the rounding error. The more computations we make, the more likely round-off errors will increase beyond a reasonable limit.

These factors suggest that we would like to get our output at a minimum cost in terms of the number of computations. The FIR filter is not always the best approach when it is important to minimize computation cycles. On the other hand, the simplicity of designing an FIR filter, combined with its inherent stability, make the FIR filter the preferred choice for many designers. (With today's high-speed processors and low-cost ICs, long delay lines are not the extreme problem they used to be.)

FIR Filter Design Methodology

As we discussed earlier in the chapter, a variety of filters can be implemented by convolving an input sequence with a transfer sequence. The trick is to come up with a transfer sequence that will produce the desired output from the actual input. While it probably is not obvious, we have already developed the tools we need to do this.

In general, the idea behind FIR filter design is to define the transfer function as a function of *frequency*. This function of frequency, generally named $H(\omega)$, is then transformed into a sequence that is a function of time: $h[n]$. The transformation is accomplished by the inverse discrete Fourier transform (IDFT). A filter is implemented by convolving $h(n)$ with the input sequence $x[n]$. The resulting sequence, $y[n]$, is the output of the filter. This process works for either a real-time process or an off-line processing system.

In practice, the sequence described above will not always produce the desired output $y[n]$. Or, more simply, the filter will not always do what we designed it to do. If this is the case, the function $H[\omega]$ or the sequence $h[n]$ will generally be tweaked to obtain the desired output. This whole design process is shown in Figure 6.3.

Technology Trade-offs

Theoretically, any realizable filter can be designed using this simple process. In some cases, however, it will turn out that no amount of tweaking will yield a

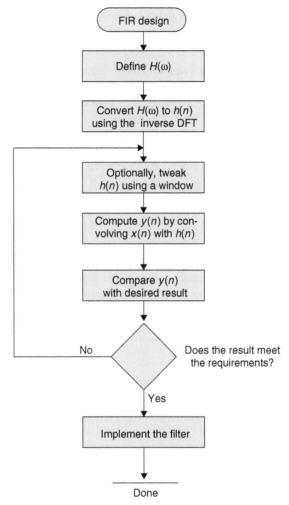

FIGURE 6.3   Filter design process for an FIR filter

practical design. As we have discussed, an FIR filter implementation may end up requiring a great number of taps. From a practical point of view, a large number of taps often leads to "mushy" or noisy filter response. When this happens, more sophisticated (that is, more complicated) filters can be tried, as discussed later.

The easiest way to understand this design method is with an example.

FIR Design Example

In this section, we will demonstrate the design of a typical DSP application. There are numerous software tools available that can be used to generate signals and design filters, ranging from sophisticated packages, like Mathworks's Matlab,

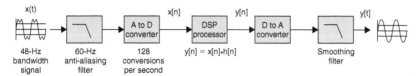

FIGURE 6.4   Block diagram for low-pass filter example

to free or shareware programs. Here we will make use of the accompanying DSP Calculator software. This example assumes a basic understanding of DSP architecture, convolution, and the discrete Fourier transform. If any of these seem confusing while working through the example, please refer to the appropriate chapters.

For our example, we will design and implement a low-pass filter, requiring the following steps:

- Create a sample waveform with the desired characteristics.
- Look at the spectrum of the sample waveform to ensure that it meets our needs.
- Design the low-pass filter.
- Generate a transfer function to realize the low-pass function.
- Test the design by convolving the transfer function with the sample waveform.

System Description

A block diagram of our system is shown in Figure 6.4. Our system is designed to monitor process signals for an industrial plant. The bandwidth of the signals is 0 Hz to 60 Hz. An anti-aliasing filter is in the front end of the system, and it ensures that any signals will be within this bandwidth.

The signal that we are interested in is a 16-Hz sine wave. Along with this signal is a separate, lower-amplitude, sine wave at 48 Hz.

Our task is to come up with a digital filter that will keep the 16-Hz signal but eliminate the 48-Hz signal.

Generating a Test Signal

Before we can modify a signal, we must first *have* a signal. Coming up with test signals that have the right characteristics to correctly exercise a DSP system is an important part of the design process. In this case, we can easily generate a test signal using the program *Fourier.* The first thing to do is create a working directory. Use the Windows File Manager to create a directory called c:\testsig. Next, open the DSP application group and double click on the icon labeled *Fourier.* Set up the following values in the appropriate boxes:

Frequency: 16
Amplitude: 1
Number of Samples: 128

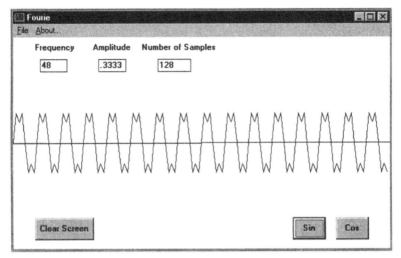

FIGURE 6.5   Sample waveform for the low-pass filter example

Then click on the Sin button. You should see a sine wave appear on the screen. Next, set the following values:

Frequency: 48
Amplitude: 0.3333
Number of Samples: 128

Then click the Sin button again. The resulting waveform should look like the one in Figure 6.5. Now save the file to `c:\testsig\x.dat`. (Use the `File/Save` command to do this.) Then close the *Fourier* window; we are done with it for now.

Now we have an input sample with the correct spectral characteristics. The next step is to prove this to be true.

Looking at the Spectrum

We can look at the spectrum of our signal using the DFT program. Double click on the DFT icon, then load in the file *c:\testsig\x.dat*. (Use the File/*Load Signal* menu to do this.) You should see the same wave that was generated in the *Fourier* program. Now click on the Transform button.

The result should look like Figure 6.6. The first thing to note is that the *x*-axis is the frequency axis. For digitally processed signals, the frequency spectrum is *always* $-\pi$ to $+\pi$. This is called the normalized frequency. Any frequency outside the range of $-\pi$ to $+\pi$ will alias to a frequency with this range. The next logical question is, of course, how does this relate to our actual frequencies?

The answer is that π corresponds to the Nyquist frequency, which is one-half of the sample *rate*. In this example, our sample rate can be assumed to be equal to the *number of samples*: 128. Therefore, the value of π corresponds to a value of 64 Hz. Our base signal is 16 Hz, which is one-fourth of 64. And that is exactly where we see the spectral peak for the 16-Hz signal: one-quarter of

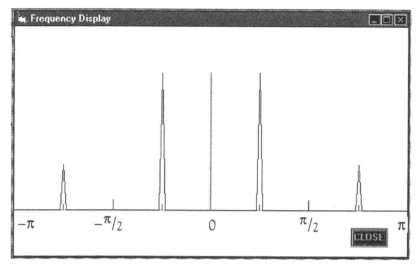

FIGURE 6.6   Resulting spectrum for the sample signal

the way from 0 Hz to π. We also see the 48-Hz spectral peak at three-quarters of the way to π.

As we would expect, the amplitude of the 48-Hz signal is one-third of the base signal's amplitude. The vertical axis does not really conform to any common physical units such as watts or volts. This is due to the way the transform works. However, the height of the spectral line can loosely be thought of as the amplitude of the signal. The vertical axis is usually scaled to conveniently show the *relative* amplitude of the signals present.

The spectrum is mirrored around the DC (that is, 0 Hz) line. The fact that the negative frequency amplitude components are an exact mirror image tells us that the input signal was either purely real or purely imaginary. Only complex signals can have a positive or negative frequency component that is not symmetrical in amplitude.

This signal meets our spectral requirements for a test signal, so we can now proceed to design and test our filter.

Design the Filter

There are a number of ways to meet our design requirements. Figure 6.7 shows one approach. We have defined a low-pass filter that simply splits the difference between the signal that we want to keep and the signal that we want to reject. So we have set our cutoff frequency at π/2, or, equivalently, 32 Hz. The filter shape is shown with a dashed line.

Filtering in the frequency domain is a simple operation: we just multiply the frequency components we want to keep by unity. All other frequency

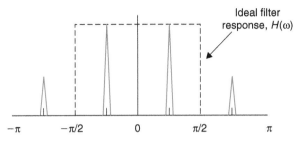

FIGURE 6.7   Desired filter shape

components are multiplied by zero. Mathematically, we can define our filter function as:

$$H(\omega) = \begin{cases} 1, -\dfrac{\pi}{2} < \omega < \dfrac{\pi}{2} \\ 0, \text{otherwise} \end{cases}$$

Theoretically, we could actually implement our filter this way. We could transform the incoming signal, zero out the 48-Hz spectral line, and then perform an inverse transform to get back to a time domain representation of the signal. Occasionally, filtering is done this way. In general, however, this is extremely inefficient from a computational and implementation point of view. Instead, we make use of the fact that multiplication in the frequency domain corresponds to convolution in the time domain:

$$H(\omega) \bullet G(\omega) \Leftrightarrow h(n) * g(n)$$

Thus, we need to generate a transfer function that, when its impulse response is transformed to the frequency domain, approximates $H(\omega)$. Notice that we said *approximates* $H(\omega)$. Our impulse function will generally not be identical to the ideal case.

We can generate a transfer function by double clicking on the Filter Design icon. Select *Filters/Low Pass* from the menu. A dialog box with two entries will come up. The first entry is the upper cutoff frequency. We decided in the last section that we wanted a value of $\pi/2$, which is equivalent to 32 Hz in this case. So, enter a value of 1.5708 for the cutoff frequency.

The number of taps determines how closely our filter will approximate the ideal $H(\omega)$. More taps will provide a closer fit. For this example, we will use 15 taps, so enter 15 into the Number of Taps box. Then click on the OK box. The cursor will change to an hourglass, indicating that the transfer function is being computed.

When the computations are done, the hourglass will turn back into the normal cursor and the transfer function will be displayed. Notice that this is the frequency domain representation of the transfer function. Save the transfer function by selecting File/Save As. Save the file to c:\testsig\h.dat. The save operation saves the *time domain* representation of the transfer function. This is also called the impulse response of the transfer function.

Feel free to experiment with the number of taps and with moving the cutoff frequency around, if you like. Close the filter design window when you are done, and we will be ready for the next step.

Convolution of the Signal

We now have our test signal, $x[n]$, and we have just generated a transfer function in the form of $h[n]$. The only thing left to do is to perform the filtering. We will do this by convolving $x[n]$ with $h[n]$.

Double click on the *Convolve* icon. Make sure that the values are set as follows:

Amplitude: 1
Number of Samples: 15

The value of 15 for the *Number of Samples* entry, in this case, corresponds to the number of taps we selected for the filter. Now select *File/Load Coefficients*. Select the file c:\testsig\h.dat. The transfer function we generated in the last step will be loaded and displayed. Notice that this is the time domain representation, so it will *not* have the same shape as shown in the filter design program.

Next set the *Number of Samples* to 128. Then select *File/Load Signal* and load the file c:\testsig\x.dat. The test signal will be displayed.

Now click on the *Convolve* button. The cursor will turn into an hourglass to indicate that the computations are being performed. A dialog box will appear when the convolution is completed. Click OK. The result of the convolution will be displayed. In this case, it is the original 16-Hz sine wave we started with.

We have successfully designed a filter that will allow the 16-Hz signal through, but will block the 48-Hz signal.

For a real application, we would take our transfer function and use it as the h values in the convolution sum:

$$y(n) = \sum_{m=-\infty}^{\infty} x[m]h[n-m]$$

Programming the DSP processor to compute this sum would then complete the process.

Windowing

The theoretical definition of the IDFT requires an infinite number of terms to transform $H(\omega)$ into $h(n)$. If we could generate, and make use of, an infinite number of terms from the IDFT we could realize the filter function perfectly. In practice, of course, we must use a finite number of terms for $h[n]$. By truncating the sequence, we are effectively distorting the original function $H[\omega]$. It turns out that we can correct for this distortion of $H[\omega]$ by applying a *compensating* distortion to $h[n]$. This process is sometimes referred to as prewarping the function $h[n]$. Generally, this prewarping process is accomplished by passing $h[n]$ through a *window function*.

Technology Trade-offs

There are a number of different windows that can be used. One of the simplest and most common is the *Bartlett window*. The Bartlett window is a simple triangular window, with straight lines for the tapering off. Many other windows exist (rectangular, Manning, and Hamming, for example), all with their distinct advantages. The rectangular window, for example, is just a truncation of the tails. The idea is the same for all windows: they are "fudge factors" that tweak the coefficients in order to achieve improved performance.

THE IIR

Infinite Impulse Response (IIR) filters use a feedback structure. If we borrow some ideas from control theory, an IIR filter can be regarded as an FIR filter inserted in a feedback loop. One of the easiest ways to approach the Infinite Impulse Response (IIR) is to start with the basic equation for the Finite Impulse Response (FIR) and then expand on this base.

If we look back at the basic FIR, we see something like this:

$$y(k) = ax[k] + bx[k-1] + cx[k-2] + \cdots + zx[k-n] \tag{6.2}$$

If we set the coefficients a, b, c … z equal to $1/(n-1)$ then we have the simple moving average filter. Or we could choose the coefficients according to some function, such as the IDFT of the frequency response of the desired filter, as we did in the last section.

FAQs

Why do we need to go beyond the FIR filter?

Theoretically, *any* filter function can be realized with Equation 6.2. What then motivates us to try something else? The answer is that while any function can be realized with an FIR, there is no guarantee that the function will be realized in an efficient manner. For example, filters with fast roll-offs take a large number of terms to implement. This has two effects: first, the filter algorithm will execute slowly and, second, the delay through the filter may be unacceptably long.

One way to improve the performance of the filter is to make use of the signal values that *have already been processed*. That is, we can make use of previous values of *y*. For example:

$$y[k] = c[0]x[k] + c[1]x[k-1] + \cdots + c[N]x[k-N] + d[1]y[k-1]$$
$$+ d[2]x[k-2] + d[2]x[k-2] + \cdots + d[M]y[k-M]$$

$$\tag{6.3}$$

Notice that we have two sets of coefficients in this form of filter function. One set is called the c coefficients and the other is the d coefficients. If we set the d coefficients equal to zero, then we have our basic FIR filter.

Equation 6.3 is often expressed more compactly as:

$$y[k] = \sum_{n=0}^{N} c[n]x[k-n] + \sum_{m=0}^{M} d[m]y[k-m] \qquad (6.4)$$

Insider Info

As a side note, the FIR filter is sometimes called a nonrecursive filter, since it does not make use of the previously processed signal. As one might expect, the IIR is sometimes called a recursive filter since it does make use of previously processed values.

The Infinite Impulse Response (IIR) filter is a little hard to get a handle on in a purely intuitive way. Unlike the FIR, which could be thought of as a modified moving average, the IIR has no convenient intuitive analog. As with the FIR, one of the major attributes of an IIR filter that we are interested in is the frequency response of the filter. In the case of the FIR, we simply took the DFT of the function we were convolving to get the frequency response. Unfortunately, this will not work on Equation 6.3 because Equation 6.3 has both the input and output terms on both sides of the equation. We need a more sophisticated tool than the DFT to handle the situation.

The answer is to make use of the z-transform. This will provide us with all the information we need. If we take the z-transform of each side of Equation 6.3 and rearrange the terms we get:

$$Y(z) - d(1)z^{-1}Y(z) - \cdots - d(M)z^{-M}Y(z)$$
$$= c(0)X(z) + c(1)z^{-1}X(z) + \cdots + c(N)z^{-N}X(z) \qquad (6.5)$$

where $Y(z)$ is the transform of the output and $X(z)$ is the transform of the input.

We can define the transfer function as the output of the filter over the input of the filter:

$$H(z) = \frac{Y(z)}{X(z)} \qquad (6.6)$$

Now, if we rearrange Equation 6.5 into the form of Equation 6.6, we get:

$$H(z) = \frac{c(0) + c(1)z^{-1} + \cdots c(N)z^{-N}}{1 - d(1)z^{-1} - \cdots - d(M)^{-M}} \qquad (6.7)$$

This is important because it shows us that the transfer function is the ratio of two polynomials in z. This means that $H[z]$ can vary quickly; the denominator can be used to drive the overall response. This rapid change in $H[z]$ is another way of saying that the filter can have very sharp transition regions, and this can be achieved with far fewer terms than would be required with an FIR filter.

The next step is to rearrange Equation 6.7 into the form of summations, where M is the number of samples we're transforming:

$$H[z] = \frac{\sum\limits_{n=0}^{N} c_n z^{-n}}{1 - \sum\limits_{m=1}^{M} d_m z^{-m}} \tag{6.8}$$

Now, just as we did with the IDFT, we can find the frequency response by letting $r = 1$ in the definition $z = re^{j\omega}$. This gives us:

$$H[e^{-j\omega}] = \frac{\sum\limits_{n=0}^{N} c_n e^{-j\omega \frac{n}{N}}}{1 - \sum\limits_{m=1}^{M} d_m e^{-j\omega \frac{m}{M}}} \tag{6.9}$$

And, just as with the z-transform, this gives us the frequency response as a complex function. It is, in fact, the value above the unit circle in the z-plane. A common practice is to take the absolute value of both sides of Equation 6.9. For simplicity, the resultant function is usually expressed as a simple function of ω:

$$H[\omega] = \left| \frac{\sum\limits_{n=0}^{N} c_n e^{-j\omega \frac{n}{N}}}{1 - \sum\limits_{m=1}^{M} d_m e^{-j\omega \frac{m}{M}}} \right| \tag{6.10}$$

Or, in other words, we can find out the frequency response of an IIR filter from its coefficients. Since the IIR filter is a ratio of polynomials, the process is more involved than is the case for the FIR filter.

So far, the discussion has followed a more or less standard textbook development. That is, the discussion assumes that we know the coefficients, and that we want to find out what filter response they will give us. The problem with this is that in the real world we generally know what frequency response we *want*. The problem is to come up with the *coefficients*.

The news on obtaining the coefficients for an IIR is mixed. The bad news is that there is no simple and practical way of analytically deriving the coefficients if we are given the desired transfer function. The good news is that there are numerous software tools that make the design of IIR filters relatively straightforward.

Conceptually, an IIR can be designed by starting off with a conventional analog filter. Normally, the filter is expressed in the Laplace form. The Laplacian of the filter is then mapped from the s-plane onto the z-plane. The coefficients of the z-plane representation are then found. This is the approach generally taught in an academic course on DSP filter design. In practice, the process is quite tedious, and not often performed by working engineers.

> **Alert!**
>
> For practical IIR design, it is generally a good idea to use one of the better filter design software packages. There are a number of reasons for this, mostly centering around the touchy behavior of the IIR. In the case of the FIR filter, we did not have to worry about filter stability, nor did we have to worry a great deal about the phase of the filter. This is not true with the IIR. It is quite possible to design an IIR that has the desired frequency response but is unusable because of stability. It is important to note that even if an IIR is technically stable, it may still exhibit an unacceptable amount of ringing or phase distortion.

With all of these caveats noted, we will now proceed to design an IIR filter. We will design it to meet the same basic requirements as the FIR filter example from the last chapter: a low-pass filter that will pass a frequency at a digital frequency of $\pi/4$, and eliminate a signal at $3\pi/4$. The reason that we are going ahead with the design of the IIR using a somewhat analytical approach needs to be addressed. While we highly recommend the use of professional-quality filter design software for developing digital filters (especially IIRs), designing an IIR from basic principles can illustrate a number of interesting and useful concepts.

Before we proceed, we should discuss the design approach we will be using. This will give us a chance to also look at some key concepts related to the z-transform. Our approach will be to place poles and zeroes appropriately around the z-plane. From the pole/zero graph we will then generate the z-transform in factored form. Next, we will evaluate the partial fraction into a standard polynomial form. From there, we will put the :z-transform in the standard form of the definition; then, we can find the coefficients of the IIR by simple inspection.

For this approach to work, we must understand some basic ideas behind the graphical representation of the transfer function in the z-plane. The z-transform of a sequence is complex, as is the function itself. Thus, a graphical representation requires four dimensions. In practice, however, we can obtain a useful graphical image if we look at the absolute value of the transfer function. The absolute value corresponds directly to the amplitude of the transfer function response. We can also find the phase by looking at the angular component, but this is of less interest at this point in the design.

We can think of the absolute value of the transfer function as a rubber membrane above the z-plane. The poles of the transfer function are created when any of the factors in the denominator go to zero. Anything divided by zero is undefined, but let's think about what happens as the denominator approaches zero. The transfer function is going to approach infinity, or, in other words, the function will "blow up." Graphically, we can think of the poles as raising up the rubber membrane to an infinite height.

The zeroes in the numerator, on the other hand, produce a value of zero for the transfer function. The zeroes will thus "tack down" the rubber membrane that represents the transfer function. As we noted above, the frequency response of the transfer function is just the z-transform evaluated around the unit circle.

One other key piece of information is required before we proceed. Let's think about what happens on the z-plane. Any point on the z-plane is defined by:

$$z = re^{j\omega}$$

where r is the distance from the origin, and ω defines the angle, relative to the positive real axis. The key concept here, however, ω is that angular frequency. We can think about the zero frequency (DC) value lying at (1,0) on the z-plane. The positive frequencies increase, in a counterclockwise direction, until we reach the point (−1,0), which corresponds to an angular value of π. The negative frequencies increase from (1,0) in a clockwise direction until we reach (−1,0).

Now, let's think about what happens when we place a pole directly on the unit circle—at an angle of $\pi/4$, for example. Assuming we start at a frequency of DC, as the frequency increases from DC, we will approach the pole. The denominator will approach 0, and the frequency response of the filter will approach infinity. The filter will blow up; in this case, when the input frequency is $\pi/4$ the output of the filter will be undefined. In practical terms, this means that even a very small input signal at $\pi/4$ (including, for example, a small amount of noise) will cause the output of the filter to try to go to an infinite value. Such a filter is unstable, and therefore probably not of much use to us. This will be true, in fact, for almost any case in which a pole lies on or outside the unit circle.

Key Concept

We now have a general rule for filter design: *All poles must lie within the unit circle.*

The corollary to this is that, as the poles move closer to the origin, the amplitude response will decrease, and the general stability of the filter will improve. In general, if we want a sharp filter with high gain we will move the poles as close to the unit circle as practical; if we want a smooth and well-behaved filter, we move the poles as close to the origin as we can get. Note that the relative position of the poles to the zeroes will have a strong effect on the shape of the response.

This all makes more sense if we look at an example. Let's recall the parameters from our FIR example. We have a test signal that is composed of:

$$y[n] = \sin\left[2\pi\,(16)\,\frac{n}{N}\right] + \frac{1}{3}\sin\left[2\pi\,(48)\,\frac{n}{N}\right]$$

where $n = 0 \ldots N - 1$, and $N = 128$. Our sample rate was specified as 128 samples/second. Thus, the digital frequency of the fundamental component of the signal is:

$$\frac{16\,\text{Hz}}{64}\pi = \frac{\pi}{4}$$

and the third harmonic's digital frequency is:

$$\frac{48\,\text{Hz}}{64}\pi = \frac{3\pi}{4}$$

Our design specification was to pass the $\pi/4$ component, and block the $3\pi/4$ component. In the case of the FIR filter, we simply split the difference; we designed a filter that would pass frequencies below $\pi/2$, and block frequencies above $\pi/2$.

For our IIR filter, we can be more specific. We can place the zeroes on the z-plane along the $3\pi/4$ radial. The zeroes will cancel the high-frequency components. To keep the low-frequency components, we will place the poles of the filter along the $\pi/4$ radial. For reasons we will discuss shortly, we will also place a pole at the origin.

Key Concept

Remember these key design rules:

- We must maintain symmetry about the x-axis. This will give us the same response for positive and negative frequencies. It will also ensure that the coefficients in the z-transform turn out to be real, and thus the coefficients of the filter will also be real. So, wherever we place a pole or zero, we will also place its complex conjugate on the z-plane.
- To ensure that the resultant filter is causal (that is, we can build a version of it that will run in real time) the order of the denominator must be greater than the order of the numerator.
- As we noted above, all poles must be inside the unit circle to ensure stability. Zeroes can be placed anywhere on the z-plane.

The question now, of course, is: Where along the radial to place the poles and zeroes? We will start our selection by making some educated guesses. First, for the poles, a reasonable starting point would be 0.5. We want a little sharper filter, however, so we will set $r = 0.6$. From experience with playing around with this kind of design, we can guess that we will need another pole to smooth out the valley caused by the other two poles. We can achieve this by setting a pole at the origin. This also accomplishes the second design requirement above: it ensures that the denominator will have a higher degree than the numerator, therefore ensuring that our filter will be causal. The zeroes are less of an issue. We can place the zeroes directly on the unit circle at the frequency that we want to suppress.

A pole/zero plot is shown in Figure 6.8.

We will develop our filter with the use of one of the standard math packages. This will give us a chance to explore the use of these tools a little, and it makes our life much simpler. In this case, we will make use of the MathCAD package from MathSoft, Inc. Notice that we are using this package to expedite dealing with the math; we are not using it as a design tool for developing the IIR. The worksheet for the IIR filter is shown in Figure 6.9.

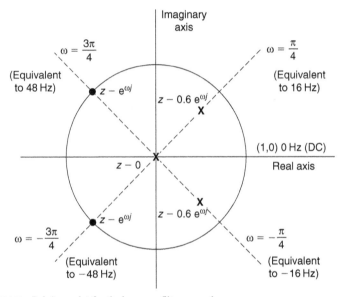

FIGURE 6.8   Pole/zero plot for the low-pass filter example

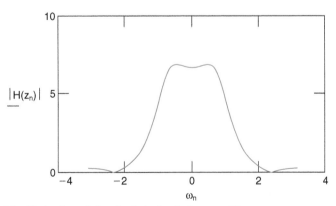

FIGURE 6.9   Electronic worksheet for designing the low-pass filter

The first thing to note in Figure 6.9 is the initial calculation that we perform at the top of the page. We have chosen a value for N of 40. This needs a little explanation. This value determines the number of points we will *plot* when we look at the frequency response of our filter. That is all it does; it is not related in any way to the sample rate or the coefficients of the filter. Next, we define an index. In this case, n will take on values from 0 to N. We need this index so that we can compute discrete values of the digital frequency. We do this next when we define ω_n, which takes on values from $-\pi$ to π. We then use ω_n to compute the values of z that we will be using in our plot.

The next step is take the poles and zeroes and turn them into the z-transform. We do this by placing the poles in the numerator and zeroes in the denominator. It is fairly straightforward to perform the symbolic computations, but we let the computer do it. First we simplify the numerator, then we simplify the denominator. Notice that we did not try to simplify the entire expression, as this would lead to an unusable and needlessly complex result. As always, we can let the computer do the work, but we cannot let it do the thinking!

At this point we graph $H(z_n)$ to see if it really is close to what we are looking for. Looking at the graph in Figure 6.9, we see that the filter indeed has the frequency response we set out to obtain.

N: = 40   N is the number of points that we will be using. This number is used for generating the graph of the frequency response. It is not related to number of samples.

n:= 0 .. N   *n* is the index variable.

$$\omega_n := 2 \cdot \pi \cdot \left(\frac{n - \dfrac{N}{2}}{N} \right)$$ Here, we are computing the indexed digital frequency from $-n$ to 71.

$z_n = e^{\omega_n \cdot j}$ Next, we compute the value of z at each of the index points.

$$H(z) := \frac{\left(z - e^{\frac{3 \cdot x}{4} \cdot j} \right) \cdot \left(z - e^{\frac{-3 \cdot x}{4} \cdot j} \right)}{\left(z - 0.6 \cdot e^{\frac{x}{4} \cdot j} \right) \cdot \left(z - 0.6 \cdot e^{\frac{-x}{4} \cdot j} \right) \cdot (z - 0)}$$ We start the derivation of the transfer function by putting the zeroes over the poles:

$$H(z) := \frac{\left(z^2 + z \cdot \sqrt{2} + 1 \right)}{\left(z - 0.6 \cdot e^{\frac{x}{4} \cdot j} \right) \cdot \left(z - 0.6 \cdot e^{\frac{-x}{4} \cdot j} \right) \cdot (z - 0)}$$ We symbolically evaluate the numerator.

$$H(z) := \frac{\left(z^2 + z \cdot \sqrt{2} + 1 \right)}{\left(z^3 - 0.6 \cdot z^2 \cdot \sqrt{2} + .36 \cdot z \right)}$$ Then evaluating the denominator symbolically gives us our transfer function in a usable form.

Now that we have the z-transform, it is time to develop the actual filter. The first thing, in this case, is to look at the gain of the filter. We did not specify a gain in our design, but in general we will want a filter that has a gain of 1 at the passband frequencies. Looking at the graph in Figure 6.9, we see that our gain is well above 5. In fact, we can find out exactly what it is if we look at Figure 6.10. In Figure 6.10 we have displayed all of the internal tables that were generated in Figure 6.9. If we look at the frequency response for $-\pi/4$(i.e, $n = 15$), we see that the frequency response (column e) is 6.063. We want to scale $H(z)$ by the reciprocal of this to give us a gain of 1 at the passband frequency.

The scaled frequency response is shown in column f of Figure 6.10. Now that we have our scale factor, we can begin to work out the values for the coefficients. The z-transform then is:

$$H(z) = \frac{1}{6.063} \times \frac{z^2 + \sqrt{2}\, z + 1}{z^3 + (0.6)\sqrt{2}\, z^2 + 0.36z}$$

which, doing the arithmetic, yields:

$$H(z) = \frac{0.165z^2 + 0.233z + 0.165}{z^3 - 0.849z^2 + 0.360z}$$

Now, we must put this in the form of the definition of the transform:

$$H(z) = \frac{0.165z^2 + 0.233z + 0.165}{z^3 - 0.849z^2 + 0.360z} \times \left(\frac{z^{-3}}{z^{-3}} \right)$$

$$= \frac{0.165z^{-1} + 0.233z^{-2} + 0.165z^{-3}}{1 - 0.849z^{-1} + 0.360z^{-2}}$$

(6.11)

One of the very nice things about the z-transform is that we can find our coefficients by simple inspection. The numerator gives us the c coefficients and the denominator provides the d coefficients. The equation for our IIR then is:

$$y[n] = 0.165x[n-1] + 0.233x[n-2] + 0.165x[n-3]$$
$$+ 0.849y[n-1] - 0.360y[n-2]$$

(6.12)

Notice that the sign of the coefficients in the denominator is inverted when we put them in the form of the equation.

Several comments are in order on this filter. First, this filter perfectly meets our requirements. That is, it passes the frequencies of $\pi/4$ with a gain of exactly 1, and it blocks signals at $3\pi/4$ completely. In practice, however, this filter is not a particularly good design. It is not very flat in the passband, and the frequency transition is not particularly sharp.

n	ω_n	z_n	$H(z_n)$	$|H(z_n)|$	$\left\|H(z_n)\cdot\dfrac{1}{6.063}\right\|$
0	−3.142	−1	−0.265	0.265	0.044
1	−2.985	−0.988−0.156j	−0.251+0.051j	0.256	0.042
2	−2.827	−0.951−0.309j	−0.208+0.09j	0.227	0.037
3	−2.67	−0.891−0.454j	−0.145+0.101j	0.177	0.029
4	−2.513	−0.809−0.588j	−0.07+0.075j	0.103	0.017
5	−2.356	−0.707−0.707j	0	0	0
6	−2.199	−0.588−0.809j	0.044−0.131j	0.138	0.023
7	−2.042	−0.454−0.891j	0.032−0.32j	0.322	0.053
8	−1.885	−0.309−0.951j	−0.075−0.561j	0.566	0.093
9	−1.728	−0.156−0.988j	−0.331−0.828j	0.892	0.147
10	−1.571	−j	−0.801−1.062j	1.331	0.219
11	−1.414	0.156−0.988j	−1.555−1.137j	1.926	0.318
12	−1.257	0.309−0.951j	−2.609−0.804j	2.731	0.45
13	−1.1	0.454−0.891j	−3.76+0.325j	3.774	0.622
14	−0.942	0.588−0.809j	−4.287+2.533j	4.979	0.821
15	−0.785	0.707−0.707j	−3.12+5.199j	6.063	1
16	−0.628	0.809−0.588j	−0.258+6.694j	6.699	1.105
17	−0.471	0.891−0.454j	2.832+6.26j	6.871	1.133
18	−0.314	0.951−0.309j	5.064+4.555j	6.811	1.123
19	−0.157	0.988−0.156j	6.292+2.345j	6.715	1.108
20	0	1	6.675	6.675	1.101
21	0.157	0.988+0.156j	6.292−2.345j	6.715	1.108
22	0.314	0.951+0.309j	5.064−4.555j	6.811	1.123
23	0.471	0.891+0.454j	2.832−6.26j	6.871	1.133
24	0.628	0.809+0.588j	−0.258−6.694j	6.699	1.105
25	0.785	0.707+0.707j	−3.12−5.199j	6.063	1
26	0.942	0.588+0.809j	−4.287−2.533j	4.979	0.821
27	1.1	0.454+0.891j	−3.76−0.325j	3.774	0.622
28	1.257	0.309+0.951j	−2.609+0.804j	2.731	0.45
29	1.414	0.156+0.988j	−1.555+1.137j	1.926	0.318
30	1.571	j	−0.801+1.062j	1.331	0.219
31	1.728	−0.156+0.988j	−0.331+0.828j	0.892	0.147
32	1.885	−0.309+0.951j	−0.075+0.561j	0.566	0.093
33	2.042	−0.454+0.891j	0.032+0.32j	0.322	0.053
34	2.199	−0.588+0.809j	0.044+0.131j	0.138	0.023
35	2.356	−0.707+0.707j	0	0	0
36	2.513	−0.809+0.588j	0.07−0.075j	0.103	0.017
37	2.67	−0.891+0.454j	−0.145−0.101j	0.177	0.029
38	2.827	−0.951+0.309j	−0.208−0.09j	0.227	0.037
39	2.985	−0.988+0.156j	−0.251−0.051j	0.256	0.042
40	3.142	−1	−0.265	0.265	0.044
(a)	(b)	(c)	(d)	(e)	(f)

FIGURE 6.10 Table of the computations used in Figure 6.9

Insider Info

We could have done much better on this filter by starting with one of the standard analog filters, and mapping the poles and zeroes onto the z-plane. Or, more practically, we could have used a good filter design software package.

Another factor that can cause problems when designing with IIR filters is that the phase of the filter is not linear. Certain frequency components may come out of the filter skewed with respect to other components. All of these factors make it important to carefully evaluate any IIR. The best approach is to use design software to generate plots of the frequency response, phase, group delays, and the pole/zero plots. Remember, the poles of an IIR are the roots of the polynomial in the denominator. The zeroes are the roots of the polynomial in the numerator.

Once the plots for a given IIR look good, it is a good idea to simulate the filter and feed in samples of actual signals. The output of the filter can then be evaluated to see if it will create any problems for the given application.

FAQs

Which to use: the FIR or the IIR?

This is a good conversational bomb to drop on a group of DSP experts! Some will argue that, due to the computational efficiency, only IIRs are of any practical use. Others will argue that, due to issues of stability, phase, etc., FIRs are the best choice, with IIRs reserved only for rare cases where the work cannot be handled by an FIR.

In practice, naturally, the decision depends upon the circumstances. FIRs may take 32, 64, 128 terms or more to accomplish a filter requirement. This number of computations may produce an unacceptable loss of precision, especially if the math is done with integers. Or it may simply be too slow. In these cases, it may well be best to go to the IIR. On the other hand, the conceptual, design, and implementation simplicity make the FIR the logical place to start on any design requirement.

INSTANT SUMMARY

In this chapter we have worked with both Finite Impulse Response (FIR) filters and Infinite Impulse Response (IIR) filters.

The FIR filters are essentially sophisticated versions of the simple moving average filter. An FIR is designed by specifying the transfer function $H(\omega)$. The function $H(\omega)$ is then converted to a sequence using the IDFT. This sequence, $h(n)$, then becomes the coefficients of the filter. The FIR is then realized by convolving the input with $h(n)$.

The FIR filter has a number of significant advantages. It is unconditionally stable, easily designed, and easily implemented. It is possible to design an FIR filter with a linear phase delay. The one major disadvantage of the FIR is that it can require a large number of computations to implement.

IIR filters are more complex and much more difficult to understand intuitively than FIR filters. We worked through a design of an IIR filter using the approach of placing poles and zeroes appropriately around the z-plane. From the pole/zero graph we then generated the z-transform in factored form and evaluated the partial fraction into a standard polynomial form. From there, we put the z-transform in the standard form of the definition and could find the coefficients of the IIR by simple inspection. The high potential performance of the IIR was noted, but we also pointed out the risks of using the IIR.

Chapter 7

Applications of DSP

In an Instant

- Definitions
- Measurement and Analysis
- Telecom
- Audio and TV
- Automotive
- Nonlinear Applications
- Instant Summary

Definitions

In this chapter we will look at applications of DSP. These can be divided into two classes. The first class consists of applications that *could* be implemented using ordinary analog techniques, but where the use of digital signal processing increases the performance considerably. The second class of applications are those that *require* the use of digital signal processing and cannot be built using entirely analog methods. DSP has caused many changes in a broad range of fields, which we will cover briefly in this chapter. Each of these fields has developed its own DSP technology, with its own pertinent techniques and algorithms.

First, we'll define some terms encountered in this chapter. *Superposition* is an important technique used in DSP systems; it breaks a complicated problem down into numerous, easier problems. It can only be used on *linear* systems, which means that a system must have two properties, *additivity* and *homogeneity*. Additivity means that if an input A produces an output x, and input B produces output y, then input A + B produces output x + y. Homogeneity means that any change in a signal's amplitude will produce an identical change in the output signal's amplitude. In this chapter, we'll also look at some *nonlinear* system applications. Some examples of nonlinear systems include circuits for peak detection, squaring, frequency doubling, etc.; multiplication of a signal by another, such as in automatic gain controls or modulation; and systems that have a threshold, such as digital logic gates. (In dealing with nonlinear systems in DSP, the main technique is to find some way to make them look like linear systems.)

MEASUREMENTS AND ANALYSIS

Digital signal processing traditionally has been very useful in the areas of measurement and analysis in two different ways. One is to precondition the measured signal by rejecting the disturbing noise and interference or to help interpret the properties of collected data by, for instance, correlation and spectral transforms. In the area of medical electronic equipment, more or less sophisticated digital filters can be found in electrocardiograph (ECG) and electroencephalogram (EEG) equipment to record the weak signals in the presence of heavy background noise and interference.

As pointed out earlier, digital signal processing has historically been used in systems dealing with seismic signals due to the limited bandwidth of these signals. Digital signal processing has also proven to be very well suited for air and space measuring applications, such as analysis of noise received from outer space by radio telescopes or analysis of satellite data. Using digital signal processing techniques for analysis of radar and sonar echoes is also of great importance in both civilian as well as military contexts.

Insider Info

DSP has caused a revolution in radar systems. It has allowed compressing of the RF pulse after it is received, filtering to reduce noise, and selecting and generating various pulse widths and shapes—all at speeds of several hundred megahertz! All of this has increased the range of radar and given better distance determination.

Another application is navigational systems. In global positioning system (GPS) receivers (RXs) today, advanced digital signal processing techniques are employed to enhance resolution and reliability.

TELECOMMUNICATIONS

Digital signal processing has basically revolutionized the telecommunications industry. It is used in many telecommunication systems today; for instance, in telephone systems for dual-tone multi-frequency (DTMF) signaling, echo canceling of telephone lines and equalizers used in high-speed telephone modems. Further, error-correcting codes are used to protect digital signals from bit errors during transmission (or storing) and different data compression algorithms are utilized to reduce the number of data bits needed to represent a given amount of information.

Digital signal processing is also used in many contexts in cellular telephone systems, for instance speech coding in mobile or global systems for mobile communication (GSM) telephones, modulators and demodulators, voice scrambling and other cryptographic devices. It is very common to find five to ten microcontrollers in a low-cost cellular telephone. An application dealing with high frequency is the directive antenna having an electronically controlled beam. By using directive antennas at the base stations in a cellular system, the base station can "point" at the mobile at all times, thereby reducing the transmitter (TX) power needed. This in turn increases the capacity of a fixed bandwidth system in terms of the number of simultaneous users per square mile, and so increases the service level and the revenue for the system operator.

> **Insider Info**
>
> *One major problem in long-distance phone communication is echo due to the time delays. DSP helps to solve this problem by measuring the returned signal and then creating an antisignal that cancels the echo. This technique is also used in speakerphones to get rid of audio feedback.*

The increased use of the Internet implies the use of digital processing in many layers, not only for signal processing in asymmetric digital subscriber loop (ADSL) and digital subscriber loop (DSL) modems, but also for error correction, data compression (images and audio) and protocol handling.

AUDIO AND TELEVISION

In most audio and video equipment today, such as DVD and CD players, digital audio tape (DAT), and MP3 players, digital signal processing is mandatory. This is also true for most modern studio equipment as well as more or less advanced synthesizers used in today's music production. Digital signal processing has also made many new noise suppression and companding systems (e.g., Dolby™) attractive.

Digital methods are not only used for producing and storing audio and video information, but also for distribution. This could be between studios and transmitters, or even directly to the end user, such as in the digital audio broadcasting (DAB) system. Digital transmission is also used for broadcasting of television (TV) signals. High definition television (HDTV) systems utilize many digital image processing techniques. *Digital image processing* can be regarded as a special branch of digital processing having many things in common with digital signal processing, but dealing mainly with two-dimensional image signals. Digital image processing can be used for many tasks, e.g., restoring distorted or blurred images, morphing, data compression by image coding, identification and analysis of pictures and photos.

> **Insider Info**
>
> *As in many other areas, DSP has solved major problems with medical equipment. A computed tomography (CT) scanner uses signals from many x-rays and stores these as digital data. Using DSP techniques, this data is used to calculate images that represent slices through the human body, which show a lot more detail than earlier techniques and allow better diagnosing and treatment. In 1979, Godfrey N. Hounsfield and Allan M. Cormack shared the Nobel Prize in Medicine for their work on CT. (Computed tomography was originally called computed axial tomography, or CAT scanning. This term is still often used by the public, but is frowned upon by medical professionals.)*

HOUSEHOLD APPLIANCES AND TOYS

In most modern dishwashers, dryers, washing machines, microwave ovens, air conditioners, toasters and so on, you are likely to find embedded microcontrollers performing miscellaneous digital processing tasks. Almost every type of algorithm can be found, ranging from simple timers to advanced fuzzy logic systems. Microprocessors executing digital signal processing algorithms can also be found in toys, such as talking dolls, speech recognition controlled gadgets and more or less advanced toy robots.

AUTOMOTIVE

In the automotive business, digital signal processing is often used for control purposes. Some examples are ignition and injection control systems, "intelligent" suspension systems, anti-skid brakes, "anti-spin" four-wheel-drive systems, climate control systems, intelligent cruise controllers and airbag controllers.

There are also systems for speech recognition and speech synthesis being tested in automobiles. Just tell the car: "Switch on the headlights" and it will, and maybe it will give the answer: "The right rear parking light is not working." New products are also systems for background noise cancellation in cars using adaptive digital filters, and radar assisted, more or less "smart" cruise controllers.

NONLINEAR APPLICATIONS

There is an almost infinite number of nonlinear signal processing applications; a few examples are the *median filter, artificial neural networks*, and *fuzzy logic*. Some of these are devices or algorithms that are quite easy to implement using DSP techniques, but would be almost impossible to build in practice using classical analog methods.

Key Concept

Many common signal processing operations are nonlinear: rectifying, quantization, power estimation, modulation, demodulation, mixing signals (frequency translation), and correlating. Filtering a signal with fixed coefficients is linear, while using an adaptive filter, having variable coefficients, can be regarding as a nonlinear operation.

Median Filter

A median filter is a nonlinear filter used for signal smoothing. It is particularly good for removing impulsive type noise from a signal. There are a number of variations of this filter, and a two-dimensional variant is often used in DSP systems to remove noise and speckles from images.

The nonlinear function of the median filter can be expressed as

$$y(n) = \text{med}\left[x(n-k), x(n-k+1), \ldots, x(n), \ldots, x(n+k-1), x(n+k)\right] \quad (7.1)$$

where $y(n)$ is the output and $x(n)$ the input signal. The filter "collects" a window containing $N = 2k + 1$ samples of the input signal and then performs the median operation on this set of samples. The median filter itself is simple and in the standard form there is only one design parameter, namely the filter length $N = 2k + 1$.

Threshold Decomposition

Analyzing combinations of linear filters using the principle of *superposition* is in many cases easier than analyzing combinations of nonlinear devices like the median filter. However, using a method called *threshold decomposition* lets us divide the analysis problem of the median filter into smaller parts. This means decomposing it into $M - 1$ *binary* signals $x^1(n) \, x^2(n), \ldots, x^{M-1}(n)$

$$x^m(n) = \begin{cases} 1 & \text{if } x(n) \geq m \\ 0 & \text{else} \end{cases} \quad (7.2)$$

Alert!

The upper index is only an index, and does not imply "raised to." Our original M-valued signal can easily be reconstructed from the binary signal by adding them together.

$$x(n) = \sum_{m=1}^{M-1} x^m(n) \quad (7.3)$$

Now, a very interesting property of a median filter is that instead of filtering the original M-valued signal, we can decompose it into $M - 1$ "channels" (Eq. 7.2) each containing a binary median filter. Then we can add the outputs of all the filters (Eq. 7.3) to obtain an M-valued output signal (see Figure 7.1).

The threshold decomposition method is not only good for analyzing purposes, but it is also of great interest for implementing median filters. A binary median filter is easy to implement, since the median operation can be replaced by a simple vote of majority. If there are more ones than zeros, the filter output should be one. This can be implemented in many different ways. For example, a binary median filter of length $N = 3$ can be implemented using simple Boolean functions

$$y(n) = x(n-1) \cap x(n) \cup x(n-1) \cap x(n+1) \cup x(n) \cap x(n+1) \quad (7.4)$$

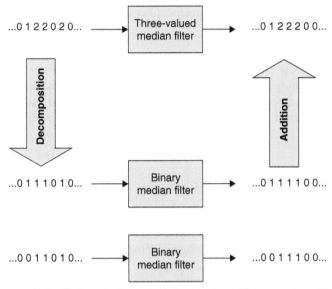

FIGURE 7.1   Median filtering of a three-valued signal by threshold decomposition, $M = 3$

where $x(n)$ and $y(n)$ are binary variables and the Boolean operations ∩ is AND and ∪ is OR. For large filter lengths, this method may result in complex Boolean calculations, and a simple counter can be used as an alternative.

If the median filter is implemented as software on a digital computer, any standard sorting algorithm like "bubblesort" or "quick sort", etc., can be used to sort the values (if a stacked filter approach is not taken). When calculating the expected sorting time, it should be noted that we do not need to sort all the values in the filter window of the median filter. We only need to continue sorting until we have found the mid-valued sample.

The use of median filters was first suggested for smoothing statistical data. This filter type has, however, found most of its applications in the area of digital image processing. An edge-preserving filter like the median filter can remove noise and speckles without blurring the picture. Removing artifacts from imperfect data acquisition, for instance horizontal stripes sometimes produced by optical scanners, is done successfully using median filters. Median filters are also used in radiographic systems, in many commercial tomographic scan systems and for processing electroencephalogram (EEG) signals and blood pressure recordings. This type of filter is also likely to be found in newer commercial digital television sets because of the very good cost-to-performance ratio.

Artificial Neural Networks

Neural networks is a somewhat ambiguous term for a large class of massively parallel computing models. The terminology in this area is quite confused in that scientific well-defined terms are sometimes mixed with trademarks and sales lingo. A few examples are: connectionist's net, artificial neural systems (ANS), parallel distributed systems (PDS), dynamical functional systems, neuromorphic systems, adaptive associative networks, neuron computers, etc.

The Models

In general, the models consist of a large number of typically nonlinear computing *nodes*, interconnected with each other via an even larger number of adaptive *links* (weights). Using more or less crude models, the underlying idea is to mimic the function of neurons and nerves in a biological brain.

"Computers" that are built using artificial neural network models have many nice features:

- The "programming" can set the weights to appropriate values. This can be done adaptively by "training" (in a similar way to teaching humans). No procedural programming language is needed. The system will learn by examples.
- The system will be able to generalize. If an earlier unknown condition occurs, the system will respond in the most sensible way based on earlier knowledge. A "conventional" computer would simply "hang" or exhibit some irrelevant action in the same situation.
- The system can handle incomplete input data and "soft" data. A conventional computer is mainly a number cruncher, requiring well-defined input figures.
- The system is massively parallel and can easily be implemented on parallel hardware, thus obtaining high processing capacity.
- The system will contain a certain amount of redundancy. If parts of the system are damaged, the system may still work, but with degraded performance ("graceful descent"). A 1-bit error in a conventional computer will in many cases "crash".

Insider Info

Neural network research has two basic motivations: one, to gain a deeper knowledge of the human brain and, two, to develop computers that can better handle abstract and not-well-defined problems, like speech recognition and face recognition.

Systems of the type outlined above have been built for many different purposes and in many different sizes during the past 50 years. Some examples are systems for classification, pattern-recognition, content addressable memories (CAM), adaptive control, forecasting, optimization and signal processing. Since appropriate hardware is still not available, most of the systems have been implemented in software on conventional sequential computers. This unfortunately implies that the true potential of the inherently, parallel artificial neural network algorithms has not been very well exploited. The systems built so far have been rather slow and small, especially when compared to the brain.

There are two classes of neural networks:

Feedforward networks: these are characterized by having separate inputs and outputs and no internal feedback signal paths. Hence, there are no stability problems. The nodes of the network are arranged in one or more discrete layers and are commonly easy to implement. Feedforward neural networks are typically used in applications like pattern recognition, pattern restoration, and classification.

Feedback networks: also known as recurrent networks, these have all outputs internally fed back to the inputs in the general case. Hence, a network of this type commonly does not have dedicated inputs and outputs. Besides the nonlinear activation function, nodes in such a network also have some kind of dynamics built in, for instance an accumulator or integrator. The main idea of this class of network is iteration. This type of network is typically used in content addressable memories (CAM) and for solving miscellaneous optimization problems.

Artificial neural networks have traditionally been implemented in two ways, either by using analog electronic circuits, or as software on traditional digital computers. In the latter case, which is relevant for DSP applications, we of course have to use the discrete time equivalents of the continuous-time expressions present in this chapter. In the case of feedforward artificial neural networks, the conversion to discrete time is trivial. For feedback networks having node dynamics, standard numerical methods like Runge–Kutta may be used.

Fuzzy Logic

A *fuzzy logic* or *fuzzy control* system performs a static, nonlinear mapping between input and output signals. It can in some respects be viewed as a special class of feedforward artificial neural networks. The idea was originally proposed in 1965 by Professor Lofti Zadeh at the University of California, but has not been used very much until the past decade. In ordinary logic, only false or true is considered, but in a fuzzy system we can also deal with intermediate levels, e.g., one statement can be 43% true and another one 89% false.

Another interesting property is that the behavior of a fuzzy system is not described using algorithms and formulas, but rather as a set of *rules* that may be expressed in natural language. Hence, this kind of system is well suited in situations where no mathematical models can be formulated or when only heuristics are available. Using this approach, practical experience can be converted into a systematic, mathematical form, in, for instance, a control system.

A simple fuzzy logic system is shown in Figure 7.2. The *fuzzifier* uses *membership functions* to convert the input signals to a form that the *inference*

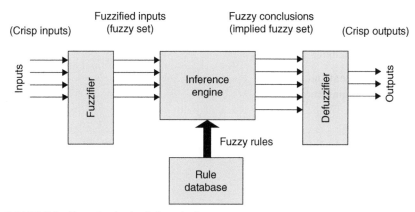

FIGURE 7.2   Example of a simple fuzzy logic system

engine can handle. The inference engine works as an "expert", interpreting the input data and making decisions based on the rules stored in the *rule database*. The rule database can be viewed as a set of "If-Then" rules. These rules can be linguistic descriptions, formulated in a way similar to the knowledge of a human expert. Finally, the *defuzzifier* converts the conclusions made by the inference engine into output signals.

Fuzzy logic is used in some signal and image processing systems and in image identification and classifying systems. Most fuzzy systems today are, however, control systems. Fuzzy regulators are suitable for applications where mathematical models are hard or impossible to formulate. The process subject to control may be time varying or strongly nonlinear, requiring elaborate theoretical work to be understood. Another suitable situation arises when there is an abundant amount of practical knowledge from manual control such that experience can be formulated in natural language rather than mathematical algorithms. It is common that people having limited knowledge in control theory find fuzzy control systems easier to understand than traditional control systems.

One drawback that exists is that there are no mathematical models available, and no computer simulations can be done. Numerous tests have to be performed in practice to prove performance and stability of a fuzzy control system under all conditions. A second drawback is that the rule database has a tendency to grow large, requiring fast processing hardware to be able to perform in real time.

Fuzzy controllers can be found not only in spaceships but also in air conditioners, refrigerators, microwave ovens, automatic gearboxes, cameras (autofocus), washing machines, copying machines, distilling equipment, industrial baking processes and many other everyday applications.

INSTANT SUMMARY

In this chapter we have summarized some of the major application areas of DSP, which include:

- Measurement and analysis
- Telecommunications
- Audio and TV
- Automotive

We also covered some nonlinear system applications that have emerged for DSP technology:

- Median filters
- Artificial neural networks
- Fuzzy logic

Chapter 8

Digital Signal Processors

In an Instant

- Definitions
- System Considerations
- Digital Signal Processors vs. Microprocessors
- The Future
- Instant Summary

Definitions

The acronym DSP is used for two terms, *digital signal processing* and *digital signal processor*. *Digital signal processing* is to perform signal processing using digital techniques with the aid of digital hardware and/or some kind of computing device. (Signal processing can, of course, be analog as well.) A specially designed digital computer or processor dedicated to signal processing applications is called a *digital signal processor*.

In this chapter, we will focus on hardware issues associated with digital signal processor chips, and we will compare the characteristics of a DSP to a conventional, general-purpose microprocessor. (The reader is assumed to be familiar with the structure and operation of a standard microprocessor.) Furthermore, software issues and some common algorithms will be discussed.

SYSTEM CONSIDERATIONS

Applications and Requirements

Signal processing systems can be divided into many different classes, depending on the demands. One way of classifying systems is to divide them into *off-line* or *batch systems* and *on-line* or *real-time systems*. In an off-line system, there is no particular demand on the data processing speed of the system, aside from the patience of the user. An example could be a data analysis system for long-term trends in thickness of the arctic ice cap. Data is collected and then stored on a data disc, for instance, and the data is then analyzed at a relatively slow

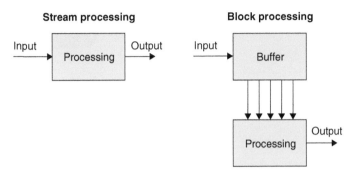

FIGURE 8.1   Stream processing and block processing

pace. In a real-time system, on the other hand, the available processing time is highly limited and must be processed quickly, in synchronism with some external process. Typical examples are digital filtering of sampled analog signals, where the filtering algorithm must be completed within the sampling period t_s. Further, in many cases no significant delay between the input signal and the output signal will be allowed. This is especially true in digital control systems, where delays may cause instability of the entire control loop (which may include heavy machinery). Most applications discussed in this book belong to the class of real-time systems, hence processing speed is crucial.

Another way of classifying signal processing systems is to distinguish between **stream data** systems and **block data** systems (see Figure 8.1). In a stream data system, a continuous flow of input data is processed, resulting in a continuous flow of output data. The digital filtering system mentioned above is a typical stream data system. At every sampling instance, data is fed to the system and after the required processing time $t_p \le t_s$ has completed, output data will be presented.

Examples of block data systems are spectrum analyzers based on fast Fourier transform (FFT) or channel decoders. In these cases, a block of data must first be inputted into the system before any computation can take place. After the processing is completed, a block of output data is obtained. A signal processing block data system often requires larger data memory than a stream data system. The most demanding applications can probably be found in the area of digital video processing. Such systems are real memory hoggers and/or require extremely high computational power. Quite often digital image processing systems are multiprocessor systems, and consist of a number of processors, dedicated subsystems and hardware. In this book, we will not consider the implementation problems associated with digital image processing systems.

Another way of classifying systems relates to the numerical resolution and the dynamic range, for instance, systems that use *fixed-point* or *floating-point* arithmetics.

Technology Trade-offs

A floating-point system often makes life easier for the designer, since the need to analyze algorithms and input data in terms of numerical truncation and overflow problems does not exist as in a fixed-point design. Floating-point arithmetic may make things easier, but as pointed out earlier, it is worth noting that a 32-bit fixed-point system, for instance, can have a higher resolution than a 32-bit floating-point system. Another point is that most systems dealing with real world signals often have some analog and/or mechanical interfacing parts. These devices have a dynamic range and/or resolution being only fractions of what a floating-point signal processing system can achieve. Hence, in most practical cases, floating-point systems are "overkill". Today, fixed-point digital processors are still less expensive and execute faster than floating-point processors.

If we try to identify the most common arithmetic operation in digital signal processing (DSP) algorithms, we will find it to be the sequential calculation of a "scalar vector product" or a convolution sum

$$y(n) = \sum_{k=a}^{b} h(k) x(n-k) = \sum_{k=a}^{b-1} h(k) x(n-k) + h(b) x(n-b) \qquad (8.1)$$

Hence, the typical operation is a repeated "multiply add accumulate (MAC)" sequence, often denoted MAC. This is found in, for instance, finite impulse response (FIR) and infinite impulse response (IIR) filter structures, where the input stream of samples $x(n)$ is convoluted with the impulse response coefficients $h(n)$ of the filter.

Key Concept

In many digital signal processing applications, we are dealing with *real-time* systems. Hence, computational speed is imperative. In particular, algorithms using the MAC-like operations should execute fast and numerical resolution and dynamic range need to be under control.

In addition, the power consumption of the hardware should preferably be low. Early DSP chips were impossible to use in battery-operated equipment, for instance mobile telephones. Besides draining the batteries in no time, the dissipated power called for large heat sinks to keep the temperature within reasonable limits. On top of this, the common commercial system requirements apply: the hardware must be reliable and easy to manufacture at low cost.

HARDWARE IMPLEMENTATION

There are mainly four different ways of implementing the required hardware:

- conventional microprocessor
- DSP chip

- bitslice or wordslice approach
- dedicated hardware, field programmable gate array (FPGA), application specific integrated circuit (ASIC).

Comparing different hardware solutions in terms of processing speed on a general level is not a trivial issue. It is not only the clock speed or instruction cycle time of a processor that determines the total processing time needed for a certain signal processing function. The bus architecture, instruction repertoire, input/output (I/O) hardware, the real-time operating system and most of all, the software algorithm used will affect the processing time to a large extent. Hence, only considering *million instructions per second (MIPS)* or *million floating-point operations per second (MFLOPS)* can be very misleading.

> **Alert!**
>
> **When trying to compare different hardware solutions in terms of speed, this should preferably be done using the actual application. If this is not possible, benchmark tests may be a solution. The ways these benchmark tests are designed and selected can of course always be subjects of discussion.**

In this chapter, the aim is not to give exact figures of processing times, nor to promote any particular chip manufacturer. ("Exact" figures would be obsolete within a few years, anyhow.) The goal is simply to give some approximate, *typical* figures of processing times for some implementation models. In this kind of real-time system, processing time translates to the maximum sampling speed and hence the maximum bandwidth of the system. Here, we have used a simple straightforward 10-tap FIR filter for benchmark discussion purposes

$$y(n) = \sum_{k=0}^{9} b_k x(n-k) \tag{8.2}$$

The first alternative is a *conventional microprocessor* system, for instance a PC-type system or some single-chip microcontroller board. By using such a system, development costs are minimum and numerous inexpensive system development tools are available. On the other hand, reliability, physical size, power consumption and cooling requirements may, however, present problems in certain applications.

Technology Trade-offs

Another problem in such a system would be the operating system. General-purpose operating systems, for instance Microsoft Windows™, are not, due to their many unpredictable interrupt sources, well suited for signal-processing tasks. A specialized *real-time operating system* (RTOS) should preferably be used. In some applications, no explicit operating system at all may be a good solution.

Implementing the FIR filter(Equation 8.2) using a standard general-purpose processor as above (no operating system overhead included) would result in a processing time of approximately $1 < t_p < 5\,\mu s$, which translates to a maximum sampling frequency, that is $f_s = 1/t_s \leq 1/t_p$ of around 200 kHz–1 MHz. This in turn implies a 100–500 kHz bandwidth of the system (using the Nyquist criterion to its limit). The 10-tap FIR filter used for benchmarking is a very simple application. Hence, when using complicated algorithms, this kind of hardware approach is only useful for systems having quite low sampling frequencies. Typical applications could be low-frequency signal processing and systems used for temperature and/or humidity control, in other words, slow control applications.

The next alternative is a *DSP chip*. DSP chips are microprocessors optimized for signal processing algorithms. They have special instructions and built-in hardware to perform the MAC operation and have architecture based on multiple buses. DSPs of today are manufactured using complementary metal oxide semiconductor (CMOS) low voltage technology, yielding low power consumption, well below 1 W. Some chips also have specially designed interfaces for external analog-to-digital (A/D) and digital-to-analog (D/A) converters. Using DSP chips requires moderate hardware design efforts. The availability of development tools is quite good, even if these tools are commonly more expensive than in the case above. Using a DSP chip, the 10-tap FIR filter (Equation 8.2) would require a processing time of approximately $t_p \approx 0.5\,\mu s$, implying a maximum sampling frequency $f_s = 2\,MHz$, or a maximum bandwidth of 1 MHz. More elaborate signal processing applications would probably use sampling frequencies of around 50 kHz, a typical sampling speed of many digital audio systems today. Hence, DSP chips are common in digital audio and telecommunication applications. They are also found in more advanced digital control systems in, for instance, aerospace and missile control equipment.

The third alternative is using *bitslice* or *wordslice* chips. In this case, we buy sub-parts of the processor, such as multipliers, sequencers, adders, shifters, address generators, etc., in chip form and design our own processor. In this way, we have full control over internal bus and memory architecture and we can define our own instruction repertoire. We, therefore, have to do all the microcoding ourselves. Building hardware this way requires great effort and is costly. A typical bitslice solution would execute our benchmark 10-tap FIR filter n i about $t_p \approx 200\,ns$. The resulting maximum sampling frequency is $f_s = 5\,MHz$ and the bandwidth 2.5 MHz. The speed improvement over DSP chips is not very exciting in this particular case, but the bitslice technique offers other advantages. For example, we are free to select the bus width of our choice and to define special instructions for special-purpose algorithms. This type of hardware is used in systems for special purposes, where power consumption, size and cost are not important factors.

The fourth alternative is to build our own system from gate level on silicon, using one or more *application specific integrated circuits (ASIC)* or *field programmable gate arrays (FPGA)*. In this case, we can design our own adders, multipliers, sequencers and so on. We are also free to use mainly any computational structure we want. However, quite often no conventional processor model is used. The processing algorithm is simply "hardwired" into the silicon. Hence, the resulting circuit cannot perform any other function. Building hardware in this way may be very costly and time consuming, depending on the development tools, skill of the designer and turn-around time of the silicon manufacturing and prototyping processes. Commonly, design tools are based on *very high-speed integrated circuit hardware description language (VHDL)* or the like. This simplifies the process of importing and reusing standard software defined hardware function blocks. Further, good simulation tools are available to aid the design and verification of the chip before it is actually implemented in silicon. This kind of software tools may cut design and verification times considerably, but many tools are expensive.

Using the ASIC approach, the silicon chip including prototypes must be produced by a chip manufacturer. This is a complicated process and may take weeks or months, which increases the development time. The FPGA on the other hand is a standard silicon chip that can be programmed in minutes by the designer, using quite simple equipment. The drawback of the FPGA is that it contains fewer circuit elements (gates) than an ASIC, which limits the complexity of the signal processing algorithm. On the other hand, more advanced FPGA chips are constantly released on the market. For instance, FPGAs containing not only matrices of programmable circuit elements, but also a number of DSP kernels are available today. Hence, the difference in complexity between FPGAs and ASICs is reduced. However, FPGAs are commonly not very well suited for large volume production, due to the programming time required.

Technology Trade-offs

There are mainly only two reasons for choosing the ASIC implementation method. Either we need the maximum processing speed, or we need the final product to be manufactured in very large numbers. In the latter case, the development cost per manufactured unit will be lower than if standard chips would have been used. An ASIC, specially designed to run the 10-tap benchmark FIR filter, is likely to reach a processing speed (today's technology) in the vicinity of $t_p \approx 2\,ns$, yielding a sampling rate of $f_s = 500\,MHz$ and a bandwidth of $250\,MHz$. Now we are approaching speeds required by radar and advanced video processing systems. Needless to say, when building such hardware in practice, many additional problems occur since we are dealing with fairly high-frequency signals.

If yet higher processing capacity is required, it is common to connect a number of processors, working in parallel in a larger system. This can be: done in different ways, either in a *single instruction multiple data (SIMD)* or in a *multiple instruction multiple data (MIMD)* structure. In an SIMD structure, all the processors are executing the same instruction but on different data streams. Such systems are sometimes also called vector processors. In an MIMD system, the processors may be executing different instructions. Common for all processor structures is however the demand for communication and synchronization between the processors. As the number of processors grows, the communication demands grow even faster.

An interesting thing is that this machine is only good at executing an appropriate type of algorithms, i.e. algorithms that can be divided into a large number of *parallel activities.* Consider our simple benchmark example, the 10-tap FIR filter. The algorithm can only be divided into 10 multiplications that can be performed simultaneously and four steps of addition (a tree of 5 groups + 2 groups + 1 group + 1 group) which has to be performed in a sequence. Hence, we will need one instruction cycle for executing the 10 multiplications (using 10 processors) and four cycles to perform the additions, thus a total of five cycles. Now, if the machine consists of 65,536 processors, each processor only has a capacity of $10^9/65,536 = 0.015$ MIPS, which is not very impressive. If we disregard communication delays, etc., we can conclude that running the 10-tap FIR filter on this "supercomputer" results in 65,526 processors out of 65,536 which are idling. The processing time will be in the range of 300 µs, in other words, considerably slower than a standard (cheaper) PC. Our benchmark problem is obviously too simple for this machine. This also illustrates the importance of "matching" the algorithm to the hardware architecture, and that MIPS alone may not be an appropriate performance measure.

DIGITAL SIGNAL PROCESSORS VERSUS MICROPROCESSORS

Conventional Microprocessors

A conventional microprocessor commonly uses a *von Neumann* architecture, which means that there is only one common system bus used for transfer of both instructions and data between the external memory chips and the processor (see Figure 8.2). The system bus consists of the three sub-buses: the data bus,

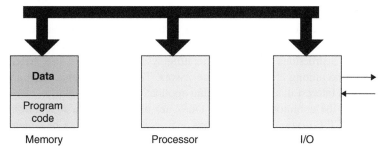

FIGURE 8.2 von Neumann architecture, program code and data share memory

the address bus and the control bus. In many cases, the same system bus is also used for I/O operations. In signal processing applications, this single bus is a bottleneck. Execution of the 10-tap FIR filter (Equation 8.2) will, for instance, require at least 60 bus cycles for instruction fetches and 40 bus cycles for data and coefficient transfers, a total of approximately 100 bus cycles. Hence, even if we are using a fast processor, the speed of the bus cycle will be a limiting factor.

One way to ease this problem is the introduction of *pipelining* techniques, which means that an *execution unit* (EU) and a *bus unit* (BU) on the processor chip work simultaneously. While one instruction is being executed in the EU the next instruction is fetched from memory by the BU and put into an instruction queue, feeding the instruction decoder. In this way, idle bus cycles are eliminated. If a jump instruction occurs in the program, a restart of the instruction queue has however to be performed, causing a delay.

Yet another improvement is to add a *cache memory* on the processor chip. A limited block (some thousand words) of the program code is read into the fast internal cache memory. In this way, instructions can be fetched from the internal cache memory at the same time as data is transferred over the external system bus. This approach may be very efficient in signal processing applications, since in many cases the entire program may fit in the cache, and no reloading is needed.

The execution unit in a conventional microprocessor may consist of an arithmetic logic unit (ALU), a multiplier, a shifter, a floating-point unit (FPU) and some data and flag registers. The ALU commonly handles 2's complement arithmetic, and the FPU uses some standard Institute of Electrical and Electronics Engineers (IEEE) floating-point formats. The *binary fractions* format discussed later in this chapter is often used in signal processing applications but is not supported by general-purpose microprocessors.

Besides program counter (PC) and stack pointer (SP), the *address unit* (AU) of a conventional microprocessor may contain a number of address and *segment* registers. There may also be an ALU for calculating addresses used in complicated addressing modes and/or handling virtual memory functions.

The instruction repertoire of many general-purpose microprocessors supports quite exotic addressing modes which are seldom used in signal processing algorithms. On the other hand, instructions for handling such things like *delay lines* or *circular buffers* in an efficient manner are rare. The MAC operation often requires a number of computer instructions, and loop counters have to be implemented in software, using general-purpose data registers.

Further, instructions aimed for operating systems and multi-task handling may be found among "higher end" processors. These instructions often are of very limited interest in signal processing applications.

Most of the common processors today are of the *complex instruction set computer* (CISC) type, i.e. instructions may occupy more than one memory word and hence require more than 1 bus cycle to fetch. Further, these instructions often require more than 1 machine cycle to execute. In many cases, *reduced instruction set computer* (RISC)-type processors may perform better in signal-processing applications.

Technology Trade-offs

In a RISC processor, no instruction occupies more than one memory word; it can be fetched in 1 bus cycle and executes in 1 machine cycle. On the other hand, many RISC instructions may be needed to perform the same function as one CISC-type instruction, but in the RISC case, you can get the required complexity only when needed.

Getting analog signals into and out of a general-purpose microprocessor often requires a lot of external hardware. Some microcontrollers have built-in A/D and D/A converters, but in most cases, these converters only have 8- or 12-bit resolution, which is not sufficient in many applications. Sometimes these converters are also quite slow. Even if there are good built-in converters, there is always need for external sample-and-hold (S/H) circuits, and (analog) anti-aliasing and reconstruction filters.

Some microprocessors have built-in high-speed serial communication circuitry, serial peripheral interface (SPI) or I²C™. In such cases we still need to have external converters, but the interface will be easier than using the traditional approach, i.e., to connect the converters in parallel to the system bus. Parallel communication will of course be faster, but the circuits needed will be more complicated and we will be stealing capacity from a common, single system bus.

The interrupt facilities found on many general-purpose processors are in many cases "overkill" for signal processing systems. In this kind of real-time application, timing is crucial and *synchronous programming* is preferred. The number of *asynchronous events*, e.g., interrupts, is kept to a minimum. Digital signal processing systems using more than a few interrupt sources are rare. One single interrupt source (be it timing or sample rate) or none is common.

Digital Signal Processors

Architecture

DSP chips often have a *Harvard*-type architecture (see Figure 8.3) or some modified version of Harvard architecture. This type of system architecture implies that there are at least two system buses, one for instruction transfers and one for data. Quite often, three system buses can be found on DSPs, one for instructions, one for data (including I/O) and one for transferring coefficients from a separate memory area or chip.

In this way, when running an FIR filter algorithm like in Equation 8.2, instructions can be fetched at the same time as data from the delay line $x(n - k)$ is fetched and as filter coefficients b_k are fetched from coefficient memory. Hence, using a DSP for the 10-tap FIR filter, only 12 bus cycles will be needed including instruction and data transfers.

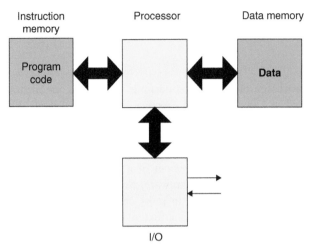

FIGURE 8.3 Harvard architecture, separate buses and memories. I/O, data and instructions can be accessed simultaneously

Many DSP chips also have internal memory areas that can be allocated as data memory, coefficient memory and/or instruction memory, or combinations of these. Pipelining is used in most DSP chips.

> **Alert!**
> Some DSP chips execute instructions in the pipeline in a parallel, "smart" fashion to increase speed. The result will in some cases be that instructions will *not* be executed in the same order as written in the program code. This may of course lead to strange behavior and cumbersome troubleshooting. One way to avoid this is to insert "dummy" instructions (for instance, no operation (NOP)) in the program code in the critical parts (consult the data sheet of the DSP chip to find out about pipeline latency). This will of course increase the execution time.

The execution unit consists of at least one (often two) arithmetic logic unit (ALU), a multiplier, a shifter, accumulators, and data and flag registers. The unit is designed with a high degree of *parallelism* in mind, hence all the ALUs, multipliers, etc., can be run simultaneously. Further, ALUs, the multiplier and accumulators are organized so that the MAC operation can be performed as efficiently as possible with the use of a minimum amount of internal data movements. Fixed-point DSPs handle 2's complement arithmetic and binary fractions format. Floating-point DSPs use floating-point formats that can be IEEE standard or some other non-standard format. In many cases, the ALUs can also handle both *wrap-around* and *saturation* arithmetic which will be discussed later in this chapter.

Many DSPs also have ready-made *look-up tables (LUT)* in memory (read only memory (ROM)). These tables may be A-law and/or μ-law for companding systems and/or sine/cosine tables for FFT or modulation purposes.

Unlike conventional processors having 16-, 32- or 64-bit bus widths, DSPs may have uncommon bus widths like 24, 48 or 56 bits, etc. The width of the instruction bus is chosen such that an RISC-like system can be achieved, i.e. every instruction only occupies one memory word and can hence be fetched in 1 bus cycle. The data buses are given a bus width that can handle a word of appropriate resolution, at the same time as extra high bits are present to keep overflow problems under control.

The address unit is complicated since it may be expected to run three address buses in parallel. There is of course a program counter and a stack pointer as in a conventional processor, but we are also likely to find a number of *index* and *pointer* registers used to generate data memory addresses. Quite often there is also one or two ALUs for calculating addresses when accessing delay lines (vectors in data memory) and coefficient tables. These pointer registers can often be *incremented* or *decremented* in a *modulo* fashion, which for instance simplifies building circular buffers. The AU may also be able to generate the specific *bit reverse* operations used when addressing butterflies in FFT algorithms.

Further, in some DSPs, the stack is implemented as a separate last in first out (LIFO) register file in silicon ("hardware stack"). Using this approach, pushing and popping on the stack will be faster, and no address bus will be used.

Instruction Repertoire

The special **multiply add accumulate (MAC)** instruction is almost mandatory in the instruction repertoire of a DSP. This single instruction performs one step in the summation of Equation 8.2, i.e., multiplies a delayed signal sample by the corresponding coefficient and adds the product to the accumulator holding the sum. Special instructions for rounding numbers are also common.

There are also a number of instructions that can be executed in parallel to use the hardware parallelism to its full extent. Further, special prefixes or postfixes can be added to achieve repetition of an instruction. This is accomplished using a special *loop counter* implemented in hardware as a special loop register. Using this register in loops, instruction fetching can be completely unnecessary in some cases.

Interface

It is common to find built-in high-speed serial communication circuitry in DSP chips. These serial ports are designed to be directly connected to coder-decoders (CODECs) and/or A/D and D/A converter chips for instance. Of course, parallel I/O can also be achieved using one of the buses.

The interrupt facilities found on DSP chips are often quite simple, with a fairly small number of interrupt inputs and priority levels.

THE FUTURE

It is not very risky to predict that coming DSP chips will be faster, more complex and cheaper. In some respects, there is also a merging of conventional microprocessor chip technology, DSPs and FPGA structures taking place. FPGAs with a number of DSP kernels on-chip are on the market, and there are more to come. Applications implemented using FPGAs are becoming more common. There have also been general-purpose microprocessor chips around for a while, having MAC instructions and other typical DSP features. New, improved simulators, compilers and other development tools are constantly being launched on the market, making life easier for the designer.

DSP chips are used in many embedded systems today in large volume consumer products like cellular mobile telephones. As the processing speed of the DSPs increases, new applications will develop continuously. One interesting area is *radio technology*. An average cellular mobile telephone today contains a number of DSP chips. Classical radio electronics circuitry occupies only a small fraction of the total printed circuit board area. As DSP chips get faster, more of the radio electronics circuits will disappear and typical radio functions like filtering, mixing, oscillating, modulation and demodulation will be implemented as DSP software rather than hardware. This is true for radios in general, not only cellular mobile telephones. For instance, radio systems based on *wideband code division multiple access (WCDMA)* and *ultra wideband (UWB)* will depend heavily on DSP technology. The technology of *software defined radio (SDR)* is growing in importance as the DSP chips get faster.

DSPs will improve but still the *programming* of DSPs will be more complex and demanding. Not only are good programming skills required, but also considerable knowledge in signal theory, numeric methods, algorithm design and mathematics.

INSTANT SUMMARY

In this chapter the following issues have been treated:

- System considerations, including technology type, speed, architectures
- Fixed and floating-point format and numerical problems
- DSP vs. conventional microprocessors
- What the future may hold.

Analog-to-Digital Converter (ADC) — Converts an analog voltage into a digital number. There are a number of different types, but the most common ones found in DSP are the Successive Approximation Register (SAR) and the Flash converter.

Analog Frequency — The analog frequency is what we normally think of as the frequency of the signal. *See Digital Frequency.*

Anti-Aliasing Filter — A filter that is used to limit the bandwidth of any incoming signal.

Digital Signal Processing (DSP) — As the term states, this is the use of digital techniques to process signals. Examples include the use of computers to filter signals, enhance music recordings, study medical and scientific phenomena, create and analyze music, and numerous other related applications.

Digital-to-Analog Converter (DAC) — Converts a digital number to an analog voltage.

Digital Frequency — The digital frequency is the analog frequency scaled by the sample interval. If λ is the digital frequency, λ is the analog frequency, and T is the sample period, then $\lambda = f/T$. The digital frequency is normally expressed over the range of $-\pi$ to π. See *Analog Frequency.*

Discrete Fourier Transform (DFT) — A computational technique for computing the transform of a signal Normally used to compute the spectrum of a signal from the time domain version of the signal. See *Inverse Discrete Fourier Transform (IDFT) Fourier Transform,* and *Fast Fourier Transform (FFT).*

DSP Processor — DSP processors are specialized to perform computations in a very fast manner Typically, they have special architectures that make moving and manipulating data more efficient. Typically, DSP processors have both hardware and software features that are optimized to perform the more common DSP functions (convolution, for example.)

Fast Fourier Transform (FFT) — Computationally efficient version of the *Discrete Fourier Transform.* The FFT is based on eliminating redundant computations often found in processing the DFT. For large transforms, the FFT may be thousands of times faster than the equivalent DFT. See *Inverse Discrete Fourier Transform (IDFT)*, *Fourier Transform,* and *Fast Fourier Transform (FFT)*.

Finite Impulse Response Filters (FIR) — A filter whose architecture guarantees that its output will eventually return to zero if the filter is excited with an impulse imput. FIR filters are unconditionally stable. See *Infinite Impulse Response Filter.*

Fourier Transform — A mathematical transform using sinusoids as the basis function. See the *Discrete Fourier Transform (DFT)* and the *Fast Fourier Transform (FFT)*.

Fourier Series — A series of sinusoid wave forms that, when added together, produce a resultant wave form.

Harvard Architecture — A common architecture for DSP processors, the Harvard architecture splits the data path and the instruction path into two separate streams. This increases the parallelism of the processor, and therefore improves the throughput. See *DSP Processors.*

Infinite Impulse Response Filters (IIR) — A filter that, once excited, may have an output for an infinite period of time. Depending upon a number of factors, an IIR may be unconditionally stable, conditionally stable, or unstable.

Inverse Discrete Fourier Transform (IDFT) — A computational technique for computing the transform of a signal. Normally used to compute the time domain representation of a signal from the spectrum of the signal. See *Discrete Fourier Transform (DFT)*, *Fourier Transform,* and *Fast Fourier Transform (FFT)*.

Linear System — A system that possesses the properties of homogeneity and additivity.

Nyquist Theorem — See *Sampling Theorem.*

Sampling Theorem — A DSP milestone, it states that to accurately reproduce a signal, we must sample at a rate greater than twice the frequency of the highest frequency component present in the signal. Also called the *Nyquist Theorem.*

Smoothing filter — A filter that is used on the output of the DAC in a DSP system. Its purpose is to smooth out the stair step pattern of the DAC's output.

Von Neumann Architecture — The standard computer architecture. A Von Neumann machine combines both data and instructions into the same processing stream. Named after mathematician Johaan Von Neumann (1903–1957), who conceived the idea.

Window — As applied to DSP, a window is a special function that shapes the transfer function. Typically used to tweak the coefficients of filters.

Burrus, C.S., Parks, T.W., *DFT/FFT and Convolution Algorithms*, John Wiley and Sons, Inc., 1985.

Foster, Caxton C., *Real Time Programming*, Addison Wesley Publishing Company, Inc., 1981.

Kester, Walt, *Mixed-Signal and DSP Design Techniques*, Elsevier Science, 2003.

Peled, Abraham, and Liu, Bede, *Digital Signal Processing*, John Wiley and Sons, Inc., 1976.

Proakis, John G., *Digital Communications*, McGraw-Hill, 1989.

Rorabaugh, C. Britton, *Digital Filter Designer's Handbook*, McGraw-Hill, 1993.

Smith, Mark J.T., and Mersereau, Russell M., *Introduction to Digital Signal Processing*, John Wiley and Sons, Inc., 1992.

Smith, Steven W., *Digital Signal Processing: A Practical Guide for Engineers and Scientists*, Elsevier Science, 2003.

Stanley, Willam D., *Network Analysis with Applications*, Reston Publishing Company, Inc., 1985.

Stearns, Samuel D., *Digital Signal Analysis*, Hayden Book Company, 1975.

Willams, Charles S., *Designing Digital Filters*, Prentice Hall, 1986.

Printed and bound by CPI Group (UK) Ltd, Croydon, CR0 4YY

03/10/2024

01040847-0012